THE COMPLETE IDIOT'S GUIDE® TO

Microbiology

Jeffrey J. Byrd, Ph.D., **and** *Tabitha M. Powledge*

ALPHA

A member of Penguin Group (USA) Inc.

ALPHA BOOKS

Published by the Penguin Group

Penguin Group (USA) Inc., 375 Hudson Street, New York, New York 10014, U.S.A.

Penguin Group (Canada), 10 Alcorn Avenue, Toronto, Ontario, Canada M4V 3B2 (a division of Pearson Penguin Canada Inc.)

Penguin Books Ltd, 80 Strand, London WC2R 0RL, England

Penguin Ireland, 25 St Stephen's Green, Dublin 2, Ireland (a division of Penguin Books Ltd)

Penguin Group (Australia), 250 Camberwell Road, Camberwell, Victoria 3124, Australia (a division of Pearson Australia Group Pty Ltd)

Penguin Books India Pvt Ltd, 11 Community Centre, Panchsheel Park, New Delhi—110 017, India

Penguin Group (NZ), cnr Airborne and Rosedale Roads, Albany, Auckland 1310, New Zealand (a division of Pearson New Zealand Ltd)

Penguin Books (South Africa) (Pty) Ltd, 24 Sturdee Avenue, Rosebank, Johannesburg 2196, South Africa

Penguin Books Ltd, Registered Offices: 80 Strand, London WC2R 0RL, England

International Standard Book Number: 1-59257-498-x
Library of Congress Catalog Card Number: 2006927518

08　07　06　　　8　7　6　5　4　3　2　1

Interpretation of the printing code: The rightmost number of the first series of numbers is the year of the book's printing; the rightmost number of the second series of numbers is the number of the book's printing. For example, a printing code of 06-1 shows that the first printing occurred in 2006.

Printed in the United States of America

Note: This publication contains the opinions and ideas of its authors. It is intended to provide helpful and informative material on the subject matter covered. It is sold with the understanding that the authors and publisher are not engaged in rendering professional services in the book. If the reader requires personal assistance or advice, a competent professional should be consulted.

The authors and publisher specifically disclaim any responsibility for any liability, loss, or risk, personal or otherwise, which is incurred as a consequence, directly or indirectly, of the use and application of any of the contents of this book.

Most Alpha books are available at special quantity discounts for bulk purchases for sales promotions, premiums, fund-raising, or educational use. Special books, or book excerpts, can also be created to fit specific needs.

For details, write: Special Markets, Alpha Books, 375 Hudson Street, New York, NY 10014.

Publisher: *Marie Butler-Knight*
Editorial Director: *Mike Sanders*
Managing Editor: *Billy Fields*
Acquisitions Editor: *Tom Stevens*
Development Editor: *Ginny Bess Munroe*
Senior Production Editor: *Janette Lynn*
Copy Editor: *Michael Dietsch*

Cartoonist: *Chris Eliopoulos*
Cover Designer: *Kurt Owens*
Book Designers: *Trina Wurst/Kurt Owens*
Indexer: *Brad Herriman*
Layout: *Brian Massey*
Proofreader: *Aaron Black*

From Jeffrey: We dedicate this book to all those who are about to enter the world of microbiology, either as students or through employment. May you come to appreciate the microorganisms presented in this book, without which life on Earth would be impossible. And to my wife Beth, and all spouses of microbiologists, who are constantly reminded of this fact.

From Tabitha: For Fred, who selflessly managed to hold my hand while doing the cooking.

Contents at a Glance

Part 1:   **Microbes and Microbiology**   1

   1   A Little about the Tiniest Life-Forms: Microbes
      and Microbiology   3
      An introduction to microbes and microbiology.

   2   Microbes 101: The Two Prokaryotes:
      Bacteria and Archaea   13
      An introduction to the bacteria and the archaea.

   3   Microbes 101, continued: The Three Eukaryotes   25
      An introduction to protozoa, algae, and fungi

   4   Viruses and Prions: Are They Alive?   39
      An introduction to the microbes that are not microorganisms.

Part 2:   **How Microbes Make a Living**   49

   5   The Molecules of Life   51
      A very brief romp through the chemistry of life.

   6   A Little About Microbial Genetics and Genomics   63
      Genetic basics.

   7   Cell Structure and Cell Structures   81
      How cells are built and what they are built of.

   8   Microbial Lifestyles   93
      How microbes eat and grow and make more microbes.

Part 3:   **Life on Earth—And on Man, Woman, and Child**   103

   9   The Social Life of Microbes   105
      Symbiosis and endosymbiosis.

   10 The Immune System: Life in the Combat Zone   119
      An introduction to the immune system.

   11 Dodging Disease by Preventing Pathogens   131
      Deal with diseases using disinfectants, antiseptics, steriliza-
      tion, and vaccination.

   12 Dealing with Disease   145
      Deal with diseases using antibiotics, antivirals, and
      antifungals.

13 Life on Man, Woman, and Child   157
    Microbes that live in and on people.

Part 4:   The Infectious Diseases   171

14 Vicious Viruses: Viral Diseases   173
    Learn about viral diseases.

15 Bad Bacteria: Bacterial Diseases   197
    Learn about bacterial diseases.

16 Pathological Protozoa and Fearsome Fungi: Eukaryote
    Microbial Pathogens   215
    What you don't know could hurt you.

17 Food Diseases and Emerging Diseases   233
    *Food poisoning, spoilage, preservation, plant and animal dis-
    eases, emerging diseases, and chronic diseases.*

Part 5:   More Life on Earth, and Under It, Too   251

18 Microbes On (and In) the Earth   253
    *Microbes in the ocean, fresh water, pollution, water treat-
    ment; microbes in soil, compost.*

19 Microbes Go to Work—and to War   265
    Biotechnology, biowar, and bioterror.

20 Microbiology in the Twenty-First Century   279
    The future of microbiology.

Appendixes

A Glossary   289

B How to Handle Microbes   305

C How to Grow Microbes   309

D Resources   313

   Index   315

Contents

Part 1: Microbes and Microbiology   **1**

1 A Little about the Tiniest Life-Forms: Microbes and Microbiology   **3**

So What Is Microbiology? 4

So What Are Microbes? 4

The Dominant Life-Forms on Earth … Really 5

Microbes and Disease 7

Microbes and Science 7

Types of Microbes 8

Eukaryotes 8

Prokaryotes 9

Evolution of Prokaryotes and Eukaryotes 9

Viruses and Prions: None of the Above 10

How Microbes Live 10

Microbes and Human Life 10

2 Microbes 101: The Two Prokaryotes: Bacteria and Archaea   **13**

Bacteria Basics 14

Life Begins with Bacteria 15

Pigeonholing Bacteria 15

Bacterial Living 17

All Together Now 19

Bad Guys, Good Guys 20

Archaea, the New Kids on the Block 22

Many Archaea Are Extreme Sports, But Not All 23

Little is Known About the Archaea 23

3 Microbes 101, Continued: The Three Eukaryotes   **25**

Birth of a Eukaryotic Cell 25

Introducing Protozoa 26

The Protozoan Lifestyle 27

Protozoa and Disease 29

Algae, Not Just Green Slime and Red Tide 30

Algae Are Plantlike Because Most Depend on Photosynthesis 30

Types of Algae 31

Algae Are a Foundation of the Marine Food Chain 32

The Bad News About Algae 33

The Fungus Among Us 34
100,000+ Species, but Only a Few Are Harmful 35
Fungi Are Master Recyclers 35
Fungal Lifestyles..36

4 Viruses and Prions: Are They Alive?   39

Viruses, the Skilled Hijackers 39
Discovery of Viruses 40
Where Did Viruses Come From? 42
Viral Diseases...42
Smallpox Success Story 43
Some Viruses Prey on Bacteria 43
Phages as Weapons Against Disease..44
Some Viruses Prey on Eukaryotes 45
Prions, Cannibals, and Mad Cows 46
Prions Fold Other Proteins into Abnormal Shapes......................47

Part 2:   How Microbes Make a Living   49

5 The Molecules of Life   51

A Brief Romp Through the Chemistry of Life 52
Linking It All Together with Chemical Bonds 52
Getting Our Feet Wet 53
The Organic Molecules in All of Us 54
Sweet Talk About Carbohydrates 55
Lipids: Fat City 56
The Big and Important Proteins...56
Nucleic Acids: DNA and RNA 58

6 A Little About Microbial Genetics and Genomics   63

Genetic Basics 64
What Are Genes?..64
What Do Genes Do? 65
How Do Genes Work? 66
So How Does DNA Make Proteins? 67
Genes Do the Two-Step 67
Transcribing Transcription 68
Translating Translation 69
Differences Between Prokaryotes and Eukaryotes.........................70

Don't Call It Junk 71
Transposons 72
Gene Sequencing: The Alphabet of Life........................73
Why Sequence Microbe Genomes? 73
Dazzling Diversity..75
Genetic Engineering of Microbes...................................76
Genetic Engineering with Microbes...............................76
Gene Therapy for People 76
Designer Genes and Making Organisms to Order.........78
Genetic Engineering Controversies 79

7 Cell Structure and Cell Structures **81**

Cells and Cell Structure 82
What Microbial Cells Have in Common 82
Membranes and Walls 84
How Many Cells Make a Microbe?...............................84
Bacterial Cell Walls and Membranes 85
Bacterial Cell Walls and Osmosis 85
Why Peptidoglycan Matters 85
Bacterial Cell Membranes 87
Archaeal Cell Walls and Membranes 89
Eukaryote Cells 90
Eukaryote Cell Walls 90
Eukaryotes Without Cell Walls....................................91

8 Microbial Lifestyles **93**

Microbes Are What They Eat..93
Eat or Be Eaten 94
How Do Microbes Eat?..94
How to Grow Microbes 95
Temperature...95
pH: The Basics of Acid and Alkaline............................96
Oxygen, or None 96
The Sex Lives, or Lack Thereof, of Microbes 97
Reproduction Without Sex—Let's Split.........................97
Fission and the Bacterial Growth Curve 97
Swapping Genes Without Sex.......................................99
Coming Together 101

Part 3:   Life on Earth—and on Man, Woman, and Child   103

9   The Social Life of Microbes   105

Symbiosis and Endosymbiosis: Living Together 106
The Feeling is Mutual 106
  Biofilm: Studying Slime *106*
  Quorum Sensing: Communication, Coordination,
  * and Competition* *108*
More Mutualism 109
  Rhizobia: Bugs in the Beans *109*
  Ruminating on Mutualism in Animals *110*
  Commensalism: Is It Real? *111*
Parasites: The Unwanted Dinner Guest 112
The Astonishing Origin of Organelles 114
  The Energy Organelles: Mitochondria and Plastids *115*
  Secondary Endosymbiosis *117*
  The Origin of the Endosymbiotic Theory of Organelle Origins *118*

10   The Immune System: Life in the Combat Zone   119

The Struggle Against Disease 120
The Innate Immune System 121
  Barrier Methods *121*
  Calling On Defenders *122*
The Adaptive (Acquired) Immune System 124
  Lymphocytes *125*
  Cell-Mediated Immunity *126*
  Humoral Immunity and Antibodies *126*
The Immune System's Memory 128
Immune System Development 129
Diseases of the Immune System 129

11   Dodging Disease by Preventing Pathogens   131

Staying Away from Pathogens...131
Getting Rid of Microbes—or at Least Keeping Them
  in Check 132
Sterilization 133
  Sterilization with Heat *134*
  Sterilization with Radiation *135*
  Sterilization with Chemicals *136*
  Sterilization with Filters *137*

Disinfectants..137
   Alcohol *138*
   Hydrogen Peroxide *138*
   Surfactants..*138*
   Heavy Metals..*138*
   Phenol and Phenolics *139*
   Have We Made the World Too Clean?....................*139*
Antiseptics 139
Vaccines: Preventing Infectious Diseases 140
   The Vaccine Success Story *141*
   Herd Immunity and the MMR Vaccine..................*142*
Other Immunological Approaches to Pathogen Prevention......143

12 Dealing with Disease **145**

Antibiotics: Weapons from the Microbe Wars..........145
   Antibiotics from Molds: Penicillin and Friends *146*
   Antibiotics from Man: Sulfa Drugs *147*
   Antibiotics from Bugs: The Mycins and Tetracyclines......*147*
   Combining Forces..*148*
   Potential Harms from Antibiotics.........................*148*
Therapies for Nonbacterial Diseases 149
   Antiviral Drugs..*149*
   Antifungal Drugs..*150*
Resistance Is Not, Apparently, Futile 150
   You Can't Keep Them Down on the Farm *152*
   Immersion in MRSA *153*
   The Difficult Clostridium....................................*153*
   What Should We Do About Antibiotic Resistance? *154*
How to Find New Antibiotics—and Why 155

13 Life on Man, Woman, and Child **157**

Normal Human Flora 158
A Gut Feeling About Microbes................................158
   The Stanford Project..*160*
   Meanwhile, In the Stomach … *161*
   The Rest of the Tract *162*
   The Straight Poop About Farting..........................*163*
Microbes on Skin 163
   Propionibacterium..*164*
   Herpesvirus *164*

The Eyes 166
Big Mouth 166
   Caring About Caries...*167*
   Gum Disease: Plenty of Periodontitis *168*
   Bad Breath: Halitosis *168*
   The Future of Dental Biofilm....................................*169*
Airing Out the Respiratory Tract 169
Urogenital Tract 170

Part 4:   The Infectious Diseases   171

14   Vicious Viruses: Viral Diseases   173
Viruses and Disease 174
Picornaviruses: Colds and More......................................175
   Rhinoviruses...*175*
   Enteroviruses *176*
Hepatovirus and Other Hepatitis Viruses 177
   Hepatitis A...*177*
   Hepatitis B...*178*
   Hepatitis C...*179*
   Hepatitis D and E *179*
Insect-Borne Viruses and Kin: Togaviruses and Flaviviruses......180
   Dengue *181*
   Yellow Fever...*182*
   West Nile Virus *182*
   Rubella..*183*
   Bunyaviruses: More Arboviruses Plus Hantavirus.......*183*
   Arenaviruses and Rodents...*183*
   Another Vector-Borne Disease: Rabies *184*
Influenza Viruses, Old and New 184
More Viruses That Cause Colds and Flu-like Illnesses 185
   Paramyxoviruses *186*
   Coronaviruses and SARS *186*
   Adenovirus: A Research Tool Causes Many Diseases *187*
   Parvoviruses Need Help..*187*
"Stomach Flu": Not Flu at All...188
   Rotavirus: the Most Important Virus You Never Heard Of........*188*
   Norwalk Virus, a.k.a. Norovirus................................*190*

Retroviruses, HIV, and Deadly AIDS 190
   Endogenous Retroviruses...*190*
   Exogenous Retroviruses, Especially HIV *191*
More Cancer-Causing Viruses 193
   Papillomaviruses...*193*
   Herpesvirus *193*
Poxviruses: A Pox Upon Them...195

15 Bacterial Diseases   197

Bacterial Pathogens and Their Hosts 197
   Staphylococcus: *MRSA and More* *199*
   Streptococcus: *Commensals and Pathogens* *201*
   Neisseria: *Gonorrhea and Meningitis* *202*
   Bacillus: *Anthrax* *204*
   Infections from Our Normal Anaerobic Flora *205*
   Clostridium: *Gangrene, Tetanus, and Botulism* *206*
   Mycoplasma *207*
   Pseudomonas: *Friend and Sometimes Opportunist* *207*
   Brucella: *Mostly Animal Diseases* *208*
   Yersinia pestis: *Plague* *208*
   Haemophilus: *Meningitis and Deafness* *209*
   Bordetella: *Whooping Cough* *210*
   Corynebacterium diptheriae: *Diptheria* *210*
   Mycobacterium: *Tuberculosis and Leprosy*..................*210*
   Chlamydia: *STD and Blindness* *211*
   Treponema: *Syphilis*..*212*
   Other Spirochaetes *213*
   Legionella: *Legionnaires' Disease* *213*
   Rickettsiae: *Rocky Mountain Spotted Fever* *213*

16 Pathological Protozoa and Fearsome Fungi: Eukaryote Microbial Pathogens   215

Protozoa and Human Disease..215
Protozoan Diseases: Malaria..216
   Malaria: The Basics *216*
   Malaria and the Immune System *217*
   Preventing and Curing Malaria....................................*219*
Protozoan Diseases: Gastroenteritis,
   a.k.a. Diarrheal Disorders 219
   Entamoeba histolytica *220*
   Giardia lamblia *221*
   Cryptosporidium *221*

Protozoan Diseases: *Trichomonas vaginalis* 221
Protozoan Diseases: More Ailments Carried by Insects 222
   Trypanosoma cruzi...*222*
   Trypanosoma brucei..*223*
   Leishmania..*224*
Protozoan Diseases: *Toxoplasma gondii* 225
The Parasitical Puppeteers: Can Parasites Control
  Your Behavior? 225
Fungi and Disease 226
   Fungal Diseases: Pneumocystis jiroveci, *formerly* P. carinii *227*
   Fungal Diseases: Candida *and Candidiasis* *227*
   Fungal Diseases: Allergies, Asthma, and Mycotoxins................*229*
   Fungal Diseases: Ringworm and Other Dermafflictions *229*
   Fungal Diseases: Aspergillosis...*230*
   Fungal Diseases: Blastomycosis *230*
   A Few Other Fungal Diseases..*231*
Algal Diseases..231

17 Food Diseases and Emerging Diseases   **233**

Foodborne Disease and Food Poisoning 234
   Campylobacter *235*
   Salmonella *236*
   Escherichia coli *236*
   Shigella *237*
   Clostridium *and Botulism*...*238*
This Food Is Rotten...238
Food Preservation 240
Plants and Animals Get Diseases, Too.......................................241
   Plant Pathogens *241*
   Animal Pathogens: The Challenges...*242*
The Emergence of Emerging Diseases 243
   Some Emerging Diseases..*244*
   Ebola and Marburg Hemorrhagic Fevers *245*
Chronic Diseases as Emerging Diseases......................................247
   Microbes and Chronic Disease..*247*
   Microbes and Cardiovascular Disease......................................*248*

Part 5:   More Life on Earth, and Under It, Too   251

18   Microbes on (and in) the Earth   253

The Water of Life 253

Microbes Go To Sea 254

Plankton, Viruses, and Sediment......................................255

Fresh Water 256

Sewage, Water, and Pollution 256

Water Treatment: To You, It's Poo, but Microbes Think

It's Yummy..257

Toward Clean Water..259

The Community of Soil..259

Soil: Don't Call It Dirt!..259

Meet the Soil Microbes 260

Biological Soil Crusts 262

Where the Soil Microbes Live 262

Compost: Fine Dining for Microbes....................................263

19   Microbes Go to Work—and to War   265

Biotechnology: Making Microbes Work 265

Fermentation: Your Basic Biotech 266

"Green" Biotech: Agriculture 268

"Red" Biotech: Medicine 270

"White" Biotech: Industry..270

The Biotech Controversies 272

A Different Kind of Warfare: Bioterror and Biowar................273

Bioweapons, Past and Future 274

What Microbes Make Good Weapons?..................................275

20   Microbiology in the Twenty-First Century   279

The Future of Microbiology 280

New Technologies, New Microbes..280

Probiotics: Edible Germs 281

CSI: Microbial Forensics..283

Bolstering the Immune Response 283

Passive Immunity..284

Active Immunity: Vaccines 285

Edible Vaccines 285

Virtual Microbial Cells...286
Custom-Made Microbes via Synthetic Biology.........................286
Microbiology as a Career..287

Appendixes

A Glossary of Terms   **289**

B How to Handle Microbes   **305**

C How to Grow Microbes   **309**

D Resources   **313**

Index   **315**

Foreword

It is often said that "seeing is believing." This was certainly true for early microbiologists like Anton von Leeuwenhoek who saw and described a universe of "animicules" (i.e., microbes) in water–pepper suspensions using a handcrafted microscope. Likewise, Elie Metchnikoff became the father of cellular immunology after observing mobile cells within the transparent larva of starfish and postulating that these cells might play a role in the protection of the host against foreign invaders. The acceptance of microscopic life advanced the idea that microbes were the cause of many human illnesses even diseases such as rabies in which the etiological agent, a virus, could not be visualized even with a microscope.

We are now able to observe the myriad types of microbes with unique shapes, sizes, and structures within their natural communities using powerful microscopes. Other advances in molecular biology provide the means of identifying microbes without growing them, which is important because we don't know how to grow most microbes, and it reveals a picture of the invisible world that is even more complex than ever imagined.

Microbes are essential to life on Earth and a basal understanding of microbiology is invaluable to everyone. People usually equate microbes with devastating human illnesses such as AIDS, some cancers, cholera, influenza, gangrene, malaria, and Dengue fever to name a few. Of equal importance is the role of beneficial microbes in warding off disease, producing foods, fuels, medicines, oxygen; and serving as the chief garbage disposals and nutrient recyclers for our planet. The importance and broad spectrum of subjects make at least a portion of microbiology of interest to nearly everyone. With such important roles in disease, food production, and the proper functioning of our planet's ecosystems, microbes will undoubtedly play a pivotal role in the future of this world and perhaps beyond.

The Complete Idiot's Guide to Microbiology provides a glimpse into the good, the bad, and the ugly of microbiology; from good wine and life-saving penicillin, to bad breath and acne, to the ugly maladies associated with vicious viruses. You will learn how our bodies fight off infection and how the microbes respond with stealth tactics to go undetected. The jargon and introductory concepts to the various areas of microbiology are provided in sufficient detail to spark the interest without bogging you down with the details. You will learn about algae, archaea, bacteria, fungi, protozoa, and viruses and how some microbes socialize with each other and interact with their host— sometimes even changing the behavior of their host. After reading this book you will see and believe that microbes, even though mostly invisible, are the most important organisms on Earth.

Charles W. Kaspar, Ph.D.

Charles W. Kaspar, Ph.D., is a Professor in the Department of Food Microbiology and Toxicology at the University of Wisconsin-Madison (www.wisc.edu). He earned his doctorate at Iowa State University in 1986. His primary areas of research are in microbial extremophiles, stress tolerance, and the human pathogen *Escherichia coli* 0157:H7, and he is an author on multiple research publications on these subjects.

Introduction

Please don't be intimidated. Microbes are just about the most enthralling, wacky subjects you can study. If you give them half a chance, we think you'll agree.

Don't worry too much about the occasional technical stuff. Microbiology is science, so there's bound to be some technical stuff. But the concepts aren't really difficult once you learn the language, and we're here to help you with that.

We define new terms the first time we use them, and often remind you what they mean when they come up again. And we've collected all the definitions in one handy place, the glossary at the back, where you can always look something up to refresh your recollection.

A splendid crew of artists has provided explanatory drawings at crucial points in the text. That's a big help in clarifying the few structures and processes that you'll need to know about.

We've stuck to scientific convention in naming specific organisms. It begins with the *Genus* name (always capitalized) followed by the *species* name (never capitalized), and the whole thing is in *italics*. The first time an organism is mentioned, it's given its whole name, but in subsequent mentions, the *Genus* is shortened to its capital letter: *G. species*.

See, that wasn't hard at all!

How to Use This Book

Who's afraid of microbes? Lots of people, even though they usually don't need to be. Who's afraid of microbiology? Not you, not with this book in front of you explaining how astounding, spellbinding, and essential microbes are.

We'll take you by the hand and give you a guided tour. We wrote it for beginners, and we don't assume any previous knowledge of this subject—although if you do know something about microbiology, we bet you'll find things here you didn't know. We'll take you by the hand and give you a guided tour of the marvels of this invisible world. Yes, there are some technical bits. But we know the lingo, so we'll translate and make it all clear.

Along the way, you'll get to know viruses and bacteria, protozoa and algae, and the fungi among us—how they live, what they do, and how they rule the world even while hidden from view.

We've divided the book into five sections for easy consumption.

Part 1, "Microbes and Microbiology," is Microbes 101. The section introduces you to the tiniest life forms—and also to the microbes that are not alive. Meet the bacteria, the viruses, the protozoa, the algae, and the fungi. Also meet the new kids on the block, the recently discovered archaea, a brand-new invisible world.

Part 2, "How Microbes Make a Living," is just what it sounds like. Here we give a quick rundown on the chemistry of life, a little about microbial genetics and genomics, cells and cell structure, and how microbes eat and grow and (blush) have sex (although most of them don't, sorry to disappoint you).

Part 3, "Life on Earth—and on Man, Woman, and Child," explores one of the most intriguing microbial traits: how sociable they are. Microbes are not solitary. They live in communities. In fact, some of them *are* communities to themselves: microbes inside microbes inside microbes. We'll tell you how the immune system works to fend microbes off—and how the wily microbes have found ways to evade it. We'll tell you about ways of preventing disease and treating it. And—perhaps most fascinating of all—we'll tell you all about those 9 out of 10 of your cells that aren't you at all, they're microbes.

Part 4, "The Infectious Diseases," is an overview. We cover diseases caused by viruses, by bacteria, by protozoa, and by fungi. You'll also learn about diseases lurking in food and water, and about emerging diseases that are worrying public health officials. And finally, the fascinating fact that several chronic diseases that we don't think of as infectious (like heart disease and cancer) can be triggered by microbes.

Part 5, "More Life on Earth, and Under It, Too," explores the microbial life in the seas, in our water, and in our soil. We don't know much about it, but we do know it's a vast, invisible universe waiting to be explored. Then we'll move on to biotechnology, a very old way of using microbes to make products that is bursting out in all kinds of new directions. We'll also cover bioweapons: what microbes might be used by bioterrorists and in future wars. Finally, we'll explore some of the new frontiers in twenty-first century microbiology: probiotics, modulating the immune system to fight disease, microbial forensics, and designer microbes from synthetic biology.

About the Extras

Sometimes we want to call your attention to a particularly important concept or a fun fact or even issue a word of caution. We do that with the help of four kinds of extras:

def•i•ni•tion

Here's where we handle some definitions. But you'll find new terms defined in the text itself, and also collected in the glossary at the back.

Tiny Tips

These contain—surprise!—tips. We think they'll help draw attention to and clarify some concepts.

Wee Warnings

These will help alert you to a confusing or controversial topic, or help make clear when some facts are still in doubt.

Little Did You Know

These extras are usually repositories of fun facts or snippets of history about microbiological science.

Acknowledgments

The Complete Idiot's Guide to Microbiology is supported on the sturdy backs and brains of thousands of microbiologists who are the very opposite of idiots. They reveal the marvels of the invisible world that all our lives depend on. They do the difficult and imaginative work we'll tell you about. They have our everlasting gratitude. This book, obviously, would not exist without them.

We also thank each other. Unlike many joint efforts, this one has been a fruitful, efficient, and amiable collaboration from beginning to end. It was an exceptionally demanding book project that required a lot of mutual encouragement. We are lucky indeed that fortune brought us together.

Many thanks also to Tom Stevens of Alpha Books and his talented crew of top-notch editors and illustrators: Ginny Munroe, Michael Dietsch, and Janette Lynn.

And to Marilyn Allen for bringing the project to us in the first place.

And to Kai (Billy) Hung for his careful reading and comments on the manuscript. And of course we thank our patient spouses, who cheered us on: Beth Byrd, Jeff's biggest supporter even when he says yes to projects that will take more time than he has available, and Fred Powledge, author of many books who knows exactly how to hearten a writing colleague who just happens to be his wife.

Special Thanks to the Technical Reviewer

The Complete Idiot's Guide to Microbiology was reviewed by an expert who double-checked the accuracy of what you'll learn here, to help us ensure that this book gives you everything you need to know about microbiology. Special thanks are extended to Kai F. Hung, Ph.D.

Kai F. Hung, Ph.D., received his B.S. in genetics from the University of Georgia, Athens, and his Ph.D. at the University of Wisconsin, Madison, also in genetics. Currently he is a post-doctoral researcher at the Food Microbiology and Toxicology program at the University of Wisconsin, Madison, where his research projects involve extremophilic microbes.

Trademarks

All terms mentioned in this book that are known to be or are suspected of being trademarks or service marks have been appropriately capitalized. Alpha Books and Penguin Group (USA) Inc. cannot attest to the accuracy of this information. Use of a term in this book should not be regarded as affecting the validity of any trademark or service mark.

Part 1

Microbes and Microbiology

This section is Microbes 101. It introduces you to the tiniest life-forms—
and also to the microbes that are not alive. Meet the bacteria, the viruses,
the protozoa, the algae, and the fungi. Also meet the new kids on the
block, the recently discovered archaea, a brand-new invisible world.

A Little about the Tiniest Life-Forms: Microbes and Microbiology

In This Chapter

- ◆ What is microbiology?
- ◆ Types of microbes, how they live, and what they do
- ◆ Microbes and disease
- ◆ Microbes and science
- ◆ Microbes and human life

Why should you learn something about microbes? Because they are essential to—well, to everything. Microbes have dominated Earth for more than 3 billion years and are the basis for all other life-forms. Microbes account for more than 60 percent of all Earth's organic matter and weigh more than 50 quadrillion metric tons. Microbes are everywhere and constitute nine out of ten cells in the human body.

There are uncounted millions of different kinds of microbes, but only a few thousand cause disease in plants and animals. Most microbes are harmless to other life and many are beneficial, even indispensable. Microbes carry out photosynthesis, which turns sunlight into food, and is the basis for Earth's breathable oxygen atmosphere and the nutrients all other organisms consume. Microbes are the master recyclers, taking in inorganic elements useless to other organisms and decomposing organic waste to make it available to other organisms. In this chapter, we meet the microbes.

So What Is Microbiology?

Microbiology is the study of microbes. What a shocker.

But this may actually be a bit of a shocker: there is no universal definition of a microbe.

A microbe is usually defined as something like "a microscopic single-cell organism," or "another word for a microorganism, which is an organism that is too small to be seen clearly with the naked eye." But here's the problem:

◆ It's true that microbes are also called microorganisms—but not all of them are organisms; that is, alive.

◆ It's true that microbes are often spoken of as microscopic—although a few can just be seen with the unaided eye.

◆ It's true that microbes are often said to be single cells—but some are multicellular.

So What Are Microbes?

A *microbe* is an agent that isn't necessarily alive, but it's a very tiny, usually invisible (to us) entity that causes something to happen, that exerts some force or effect. People are interested in microbes chiefly because of their effects—and fascinated because these effects emerge from a seemingly invisible world.

That invisible world is largely unknown, although we are learning more about it every day.

The standard list of microbes includes bacteria, viruses, protozoa, algae, and fungi. We'll describe them all, briefly in the following chapters and in more detail in the rest of the book. For us, "microbe" is a kind of umbrella term that takes in all of the above and doesn't get a whole lot more precise than that.

For the purposes of this book, we speak as if most microbes are microorganisms—microscopic living creatures. But, speaking strictly numerically, viruses are far more plentiful than any other microbes, and they're (probably) not alive at all. Just so you know.

The Dominant Life-Forms on Earth ... Really

Hard to believe—and a little spooky—that things we can't see run the world. Microbes have dominated this planet, which has been around for 4.5 billion years, for most of its history. That means it's been a microbial world for well over 3 billion years. Yes, we said billion.

The life-forms that people think about when they hear the term organism—dogs and butterflies and geraniums and bananas—are just a tiny proportion of the diversity of life. For starters, microbes are the ancestors of all living things. They are also the planet's life-support system. None of us—or any other life-form—would be here except for microbes, and none of us could continue to be here without them either. Life without the rest of us is possible. Life without microbes is not.

Yes, some microbes cause disease, misery, and death, and some have even changed human history. However, pathogens of plants and animals probably number only a few thousand out of the many millions of different kinds of microbes. Most microbes are harmless to people. A number are beneficial, even essential.

A microbiologist at work. In 2005, Terrence Tumpey, a microbiologist at the U.S. National Center for Infectious Diseases, re-created the 1918 influenza virus in order to find out why it was so deadly. He is shown in a Biosafety Level 3 lab and is working beneath a flow hood, which filters pathogens out of the air.

(Courtesy of the CDC, James Gathany.)

Microbes account for more than 60 percent of the planet's biomass—that is, 60 percent of the total weight of all its organic matter. Yet this is still a vast unknown universe.

Consider: in 1976, the Viking Lander set out to look for microbes on Mars. At that time, more than 99 percent of the microbes in Earth's seas were unknown. Things are a little better now; perhaps only 95 percent of marine microbes are unknown today.

> ### Little Did You Know
>
> Try to get your brain around these statistics: scientists say there are 5×10^{31} microbial cells on Earth. That's 5 followed by 31 zeroes. And they weigh more than 50 quadrillion metric tons.

Consider: scientists have described more than half a million insect species—but only a few thousand of the microbes called prokaryotes have been described. They believe there are millions of prokaryote species.

Consider: if you squeezed one of those half-million insects and examined the remains in a microscope, you would find hundreds—maybe thousands—of distinct microbe species.

So what are all these microbes up to? Here are some things:

Microbes are the master recyclers. They can take in nutrients and inorganic elements that other life-forms cannot use. Thus, microbes are the base of many food chains, converting substances that are useless to other organisms into nutrients that sustain them. Microbes decompose organic waste, recycling their components back into circulation where they can be taken up by other organisms.

To be specific:

- Microbes cycle and recycle the elements life is based on: carbon, nitrogen, sulfur, hydrogen, oxygen.

- Recycling these elements in soils, microbes control soil fertility and with it all plant and animal life.

- Microbes recycle gases in the atmosphere, including the ones that create the greenhouse effect. That makes life possible on the planet, but the steady buildup of these gases is also warming it, changing the climate and forcing living things to adapt or vanish.

Microbes are everywhere. With their diversity of weird (to us) metabolisms, they can survive anywhere, and do. Microbes dwell in mines deep beneath the earth and in sediments deep beneath the ocean. They live without oxygen, in boiling hot springs, and below Antarctic glaciers.

Microbes may be everywhere, but that doesn't mean that all microbes are everywhere. Many, quite sensibly, gather where their hosts gather and are not usually found elsewhere. Even free-living microbes are more plentiful and variable in some places than in others, depending on local environmental conditions and probably other factors still to be discovered.

Microbes and Disease

When pathogens cause disease, they are simply playing out an ecological tactic in a particular habitat. They are using the host organism to make copies of themselves and transmit them to other hosts. The diseases they cause are a by-product of a pathogen doing its reproductive thing.

Very often, in fact, damage to the body is not the result of direct action by the invading microbe at all. It's the result of an immune system overreaction to infection, one so strong that it injures host tissues and organs. We talk about this phenomenon more in later chapters, especially in Chapter 10.

Tiny Tips

Once more, for emphasis: most of the millions and millions of microbes are harmless and many are beneficial. Only a few thousand are pathogens—that is, cause disease.

Microbes actually keep creatures healthy. Experimental animals that are bred to be germ-free are much less healthy than normal germy animals, and when kindly researchers allow microbes to colonize the germ-free creatures, they get healthier.

Nine out of ten cells in your body are not you at all; they are microbes. Or perhaps it would be more accurate to say that you are very nearly all microbes. Microbes do much of your digesting for you. They keep you healthy, making proteins and vitamins that you can't live without but can't make for yourself. We'll take up that mind-boggling topic in Chapter 13.

Microbes and Science

Microbes are crucial for scientific research. They are simple in structure and metabolism compared with plants and animals. They have short generation times. They are easy (some of them, anyway) to keep and grow in the lab. Increasingly, scientists study entire microbial communities and ecosystems.

So microbes are wonderful model systems for studying basic questions about the nature of metabolic pathways, about the growth and death of cells, about the origin of sex and species, about how cells communicate, about genetics, biochemistry, and many other fields of science. Microbes help scientists figure out how a particular gene causes a particular trait—and how to apply that knowledge more broadly to ourselves.

The study of how microbes are related to each other led the way to showing that all life on Earth is related.

Types of Microbes

There are uncountable kinds of microbes. But the more they studied them, the more microbiologists have been able to divide them into groups. This makes it easier to think about them, to investigate them, and to see the relationships among them.

There are three main groups of microbes. Two groups are living organisms: prokaryotes and eukaryotes. One group is probably not: viruses and prions. (We introduce some technical terms here, and define these terms in detail in later chapters.)

Eukaryotes

Eukaryotes (pronounced you-CARE-ee-oats) are organisms that contain the genetic material DNA, organized into a distinct cell structure, called a nucleus, that is surrounded by a membrane. This class includes all organisms, living and extinct, except those discussed below. Therefore among the eukaryotes are pandas, passenger pigeons, petunias, and people, as well as some kinds of microbes.

There are three main groups of eukaryotic microbes: protozoa, algae, and fungi:

◆ Protozoa (pronounced PRO-tow-ZO-ah) are simple, one-celled eukaryotes. The term *protozoa* is used largely for convenience and does not indicate close evolutionary relationships among them. Scientists lumped them together because they are not easy to classify as either animal or plant, even though some are a bit like animals and others are sort of similar to plants.

◆ Algae (pronounced AL-gee) are like plants. Only two kinds of algae are microbes and consist of just one cell; there are other algae that aren't. The microbial green algae are likely ancestors of land plants. Red algae live in the ocean; most are seaweeds except for the microbial phytoplankton, crucial to the marine food chain. Like green plants, algae get their energy via *photosynthesis*.

◆ Fungi (pronounced FUN-ji; the singular is fungus) are familiar to you in the form of mushrooms, but most fungi don't resemble mushrooms at all. Many fungi are microbes, most notably yeast, mildews, and molds.

def•i•ni•tion

Photosynthesis uses energy from light—generally sunlight—to build carbohydrates out of carbon dioxide and water. A by-product of this process is free oxygen released into the atmosphere. Almost all organisms depend on photosynthesis to provide their food, directly or indirectly. Microbes carry out more photosynthesis than green plants do.

Prokaryotes

Prokaryotes (pronounced pro-CARE-ee-oats) are organisms that contain genetic material, but this material (usually DNA) is not organized into a specific cell structure called a nucleus. There are two kinds of prokaryotes: archaea and bacteria.

Bacteria (pronounced bac-TEER-ee-ah, singular bacterium) are the microbes we know the most about. Scientists have studied them the longest and hardest, largely because these microbes cause a lot of illnesses. One type, the cyanobacteria, were the original photosynthesizer.

The archaea (pronounced are-KEE-ah) were discovered only in the 1970s, when researchers realized that genetic material and proteins in some prokaryotes were different from both eukaryotes and bacteria. Their cell structure and metabolism resemble bacteria. But the way they use their genes to make new proteins is more like eukaryotes.

Evolution of Prokaryotes and Eukaryotes

Earth is about 4 billion years old. The prokaryotes came first. Prokaryotes had Earth all to themselves for a very long time—well over a billion years. They didn't waste that time, either. They used it to devise photosynthesis, which led eventually to air that we could all breathe. We explain this in more detail in Chapters 2 and 3.

Eukaryotes didn't appear until about 2 billion years ago. This step appears to have come about because a few prokaryotes decided to live together. We take up this topic, symbiosis, in Chapter 9. The appearance of eukaryotes was a very big deal, because it made possible the evolution of more complex multicellular creatures—including us.

These two microbial cell types, eukaryote and prokaryote, may appear quite different from each other. But they possess an underlying unity. They are made of the same chemicals, and their biochemistry is similar. Their genetic material, mostly DNA, is either identical or nearly so, and the way the information in it is expressed is the same. Their essential metabolic pathways are the same, too.

Viruses and Prions: None of the Above

This group of microbes includes only viruses (which are probably not alive) and prions (which are definitely not alive).

A *virus* (pronounced VY-russ) is an infectious particle that can grow and reproduce *only* inside a host cell. Viruses are no more than dabs of genetic material enclosed in little protein jackets. Because viruses are not truly cells and must make a living by hijacking the machinery of the cells they infect, most scientists would say viruses are not alive. Even though they are not alive, viruses cause a lot of diseases, of course.

Prions (pronounced PREE-ons) cause only a few diseases, all fatal. The most famous is mad cow disease. Prions are proteins that cause other proteins to fold into abnormal shapes. They are not alive.

How Microbes Live

Microbes are divided broadly into anaerobes and aerobes. Anaerobes (pronounced ANN-air-obes) either cannot tolerate oxygen or simply ignore it. Aerobes (pronounced AIR-obes) must have oxygen to survive. Almost all eukaryotes are aerobes.

Another way of thinking about microbes is to classify them by how they get their nutrients and energy. Microbes are either autotrophs or heterotrophs.

Autotrophs (pronounced AUTO-trofs) do not eat. They make their own food from light (the photosynthesizers) or chemicals. Autotrophs are sometimes called producers.

Heterotrophs (pronounced HET-ur-o-trofs) eat autotrophs or each other. Heterotrophs are sometimes called consumers. One kind of heterotroph, the fungi, are sometimes called decomposers because they eat dead stuff, breaking it down into its component parts for recycling into other life.

Little Did You Know

The first organisms were probably heterotrophic anaerobic prokaryotes. The first autotrophs were probably anaerobes that used heat energy from hot springs to convert inorganic substances into organic ones. The first photosynthetic autotrophs were probably anaerobic bacteria.

Microbes and Human Life

Infectious disease has had enormous effects on Earth's residents, shaping their population structure and—in the case of humanity—their history. Diseases like smallpox and bubonic plague and even simple diarrhea have killed millions and not infrequently determined the outcomes of battles.

The diseases Europeans brought with them when they invaded the Western Hemisphere, and then transmitted to Native Americans, may have had more to do with their triumph than any other factor. In other places in the world, indigenous diseases may have kept potential colonizers away. We discuss a few examples of microbes' impacts on history throughout the book.

A microbe at work. The can contains a concoction of cellulose, sulfur, and the bacterium Thiobacillus thioxidans. *The goo will be painted on radioactively contaminated walls and ceilings at a UK nuclear reactor to help clean it up.* T. thioxidans *is one of the few microbes that can break down concrete. It gobbles the sulfur, forming sulfuric acid, which eats away at the contaminated concrete.*

(Courtesy of the U.S. Department of Energy.)

In addition, infectious diseases sculpt the genes of a population. The survivors often possess genetic traits that helped them resist the pathogen, and thus their descendants will be more resistant in the future.

Our fossil fuels, coal and oil, are not former dinosaurs after all. They are former microbes, mostly bacteria, that perform the final steps in breaking down organics from dead organisms. Scientists say the most abundant biomolecules on Earth are hopanoids, which are found in bacterial cell walls and also in the organic precursor of petroleum.

And when our microbial fossil fuel runs out, where will we turn? To microbes again—for example the ones that convert huge vats full of cornstalks into fuel ethanol.

Microbes are essential tools in industry and medicine. They make antibiotics, enzymes, and other chemicals that contribute to quality of life and even save it. They can produce new materials like biodegradable plastics and food that resists disease and has extended shelf life. Microbiology testing also helps keep food safe to eat.

The pharmaceutical industry uses microbes to discover new drugs and to produce them. Microbes clean up hazardous waste, decontaminate soil, and purify water. They can be weapons of war or bioterror—as we saw in the anthrax attacks through the mail that happened shortly after 9/11. But microbes can also defend against those evils.

Come along, and we'll show you.

The Least You Need to Know

- Microbiology is the study of microbes, very tiny agents that exert some force or effect on organisms.

- Some microbes are eukaryotes, which possess a nucleus surrounded by a membrane and containing DNA; these are protozoa, algae, and fungi.

- Some microbes are prokaryotes, whose genetic material is not organized into a nucleus; these are bacteria and archaea.

- Some microbes are not alive and therefore are neither prokaryotes or eukaryotes; these are viruses and prions.

- Microbes have dominated Earth for more than 3 billion years and are the basis for all other life-forms.

- There are uncounted millions of different kinds of microbes, but only a few thousand cause disease in plants and animals. Most microbes are harmless to other life and many are beneficial, even essential.

Microbes 101: The Two Prokaryotes: Bacteria and Archaea

In This Chapter

- A bit about bacteria
- Where life (and oxygen) came from
- Bacterial communities
- Bacterial black hats and white hats
- The mysterious archaea

In this chapter, we introduce the two prokaryotes: bacteria and archaea. You may think you already know quite a bit about bacteria because they are responsible for many diseases—not just ours, but diseases of plants and other animals, too. Some of them are indeed bad guys, the germs that your mom warned you about. However, most are at least harmless, and many are helpful to other life.

Archaea appear to be quite harmless, but they are more of a mystery, even to scientists. They were discovered only 30 years ago, and knowledge about them is still pretty rudimentary.

Bacteria Basics

Bacteria are the microbes we know the most about. Scientists have studied them the longest and hardest, largely because these microbes cause a lot of illness. Fewer than 1 percent of bacteria can be cultured in a lab for study. That gives you an idea how vast our ignorance is, and how much work remains to be done.

Bacteria (and perhaps archaea, which we'll talk about in a moment) are the oldest kinds of life on Earth. Bacteria are also the most numerous. More than 100 million of them reside in a spoonful of topsoil, and 100,000 occupy each square centimeter of your skin.

A schematic drawing of a "typical" bacterium, showing some of its structures.

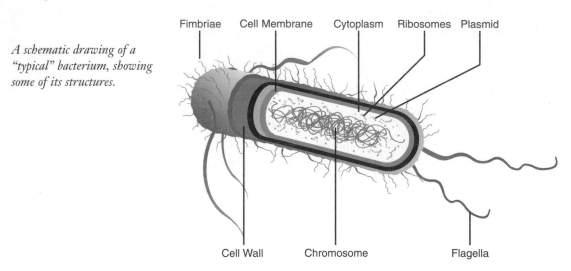

Fimbriae   Cell Membrane   Cytoplasm   Ribosomes   Plasmid

Cell Wall   Chromosome   Flagella

Bacteria is a plural term. If you have only one such microbe, you have a bacterium. In the real world, though, you rarely have just one bacterium. You may run across the term *eubacteria*, which just means true bacteria, as distinguished from the archaea. By the way, scientists often call bacteria "bugs."

Life Begins with Bacteria

The rest of life on Earth, including people, would never have happened without bacteria. Some 3 billion years ago, cyanobacteria originated photosynthesis, the process that turns light into food that (directly or indirectly) feeds almost all other life.

Cyanobacteria also produced free oxygen on what had been an oxygen-free planet. That eventually created an atmosphere that many other organisms could breathe. We'll talk more about photosynthesis in Chapter 3.

> **Tiny Tips**
>
> Photosynthesis is usually associated with algae and land plants, but it originated with the cyanobacteria, which live in water and are often wrongly called blue-green algae.

One important way to classify bacteria is by describing their relationship with oxygen:

- Aerobic bacteria can survive only in the presence of oxygen.

- Anaerobic bacteria cannot tolerate oxygen; examples are bacteria that live in ocean sediments.

- Facultative anaerobes prefer oxygen, but can live without it.

- Microaerophiles grow best in low oxygen.

- Aerotolerant anaerobes put up with the presence of oxygen but don't use it.

Two different types of bacteria using very different oxygen-handling systems can live next to each other. In short, bacteria can survive just about everywhere.

Pigeonholing Bacteria

How bacteria deal with oxygen is just one way of classifying them. As we learned in Chapter 1, sometimes bacteria (and other microbes) also are classified by the way they get energy—that is, whether they are heterotrophs or autotrophs.

Bacteria are often divided into gram-positive and gram-negative. Gram-positive bacteria have thick cell walls that display a stain used by the Danish scientist Hans Christian Gram (1853–1938). Gram-negative bacteria have thin cell walls with different chemical properties and do not show the stain. Gram staining is a quicker way of determining the presence of an infectious organism than by making a culture. We'll explain why in Chapter 7.

Sometimes scientists classify bacteria by shape. Bacteria are often described as being small and simple, so you might expect that they all look alike. Not so. Bacteria come in many different shapes. Here are the most common:

◆ Bacilli are shaped like rods or sticks.

◆ Cocci are spherical.

◆ Spirilla are squiggly but rigid like a spiral or helix; when bendable, they are called spirochetes.

◆ Vibrios are rod-shaped with a curve and look like fat commas.

Even within the shape classifications, there are a number of variations. Actinomycetes are gram-positive bacteria that trail long branching filaments, and so resemble fungi. The rod-shaped bacilli are sometimes so short and fat they look almost like cocci. A few bacteria are even flat. Others can alter their shapes.

Certain congregating patterns are characteristic of particular species as well. For example, the round cocci can divide to form long chains (*Streptococcus*) or irregular bunches like grapes (*Staphylococcus*). They can also gather in pairs (*Neisseria*), or groups of four (*Micrococcus*) or eight (*Sarcina*). *Bacillus megaterium* forms long chains.

There's a lot of variation in size, too, although bacteria mostly range from 0.2 to 10.0 µ*m*.

def•i•ni•tion

A **µm** is a millionth of a meter, usually called a micron and sometimes a micrometer. It's also a thousandth of a millimeter; 10,000 Angstrom units; 39 millionths of an inch (0.000039 inches, or 1/25,400 inches); or 1,000 nanometers (abbreviated *nm*; 1 nanometer is 0.000000001 meters). For comparison, a grain of salt is about 60 µm; one of your body cells is about 50 µm; and a human hair, about 100 µm thick. That funny-looking backward u, µ, is the Greek letter *mu*. In science and engineering, µ is pronounced micro and means one millionth.

Escherichia coli is often cited as an average bacterium: about 1.5 µm wide and no more than 6.0 µm long. The largest bacterium so far discovered is an ocean bacterium, *Thiomargarita namibiensis*. It's huge, up to 750 µm, and can even be seen without a microscope. *T. namibiensis* is about as big as the period at the end of this sentence.

Mycoplasmas, some of which infect the human respiratory and urogenital tracts, are usually said to be the smallest bacteria; some of the 90 or so species are smaller, about 150 nm wide, than the largest viruses. Some scientists think there may be even smaller bacteria, nanobacteria, believed to cause calcification of coronary arteries and other body structures. It is said they are as tiny as 100 nm wide. These claims are disputed by others who doubt that there are such living organisms.

Although there is no official classification of bacteria, the International Committee on the Systematics of Prokaryotes (ICSP) comes close. It decides on new bug names and publishes a journal on taxonomy.

The standard reference work on classification of prokaryotes is commonly known as *Bergey's Manual* and it was first published in 1923. Historically, *Bergey's Manual* has classified bacteria on the basis of their similarity to each other in form and structure. But the most recent edition, being published in five volumes, is based on genetics. Its classification system focuses on prokaryotes' evolutionary relationships with each other.

There are several ways of identifying bacteria. That task can be critical in disease because even closely related bacteria can differ greatly in pathogenicity. Bacteria with a characteristic appearance can be identified just by looking. There are also many different kinds of biochemical tests that can tell bacteria apart. More recently, researchers have been able to distinguish among bacteria by looking at their genes.

Clinical laboratories specialize in identifying microbes, and many kinds of testing procedures exist for the common ones. The task can be complex. Keeping up with microbial classification is a demanding job. Here are some reasons why:

◆ A single microbe can have more than one name.

◆ Scientists often have discovered a new bug or new strains that have developed new resistance to antibiotics, so the information is out of date—sometimes even before it has been published.

Bacterial Living

A bacterium consists of just one cell surrounded by a membrane and sometimes a gooey protective capsule. As we mentioned in Chapter 1, all bacteria are prokaryotes, meaning they don't keep their genetic material in a cell nucleus surrounded by a membrane. (That genetic material is DNA, which we'll talk about in more detail in our genetics chapter, Chapter 6.) If it was unfolded and spread out, bacterial DNA would be about a millimeter long. Instead this DNA is compressed into an area of the bacterial cell called a nucleoid (or nuclear body) and organized into one big circular structure.

Bacteria live everywhere on Earth: in soil, water, and you. We devote Chapter 13 to the subject of life on man (and woman and child). Bacteria are even residing on (and eating!) the wax statues of dancers sculpted by Edgar Degas. Many bacteria are extremophiles, which means they can live in unfriendly environments, very hot or very cold, very acid or very alkaline.

Turns out bacteria can even survive on the moon. In 1967, the United States sent the unmanned probe Surveyor 3 to the moon. In 1969, Apollo 12 astronauts recovered the probe's camera and brought it back to Earth. Conditions for recovery were completely sterile, but when NASA scientists examined the camera, they found *Streptococcus mitis*, a harmless bug that often lives in the human nose and throat. *S. mitis* had apparently gone to the moon and endured there for well over two years.

Most bacteria are free-living, but some members of *Chlamydia* and *Rickettsia* must live within the cells of other creatures. Some free-living bacteria are confirmed couch potatoes that never go anywhere. Others create slime and ooze over its slippery surfaces or get around with the help of flagella, tail-like appendages that help microbes move around. Some possess hairlike exterior structures called pili that help them stick to surfaces.

Bacteria developed one special trick that has helped them survive and prosper and spread through billions of years of Earth's tumultuous history. When frozen or dried out, they can go dormant. In this state of suspended animation, sometimes they can last for thousands of years.

Bacteria reproduce by simple cell division, called binary fission. They do exchange genetic material with other bacteria as well as other organisms, although this is not sex in the way we think of sex.

Some bacteria possess plasmids, tiny free-floating circular pieces of DNA that are separate from the bug's genome and that can make copies of themselves in the host cell. Plasmids can be transferred to other organisms fairly easily, which makes them enormously important for two reasons.

First, plasmids can be used as vectors in genetic engineering. Scientists insert the gene or genes of interest into plasmids and then transfer the plasmids to another organism. (*Vector* is one of those confusing terms, the meaning of which depends on the topic under discussion. In microbiology, a vector is genetic material that's used as a kind of vehicle for transporting other genetic material into a host cell.)

Second, plasmids can carry genes for antibiotic resistance. Because bacteria often exchange plasmids easily, it is not very difficult to pass along traits that have made some diseases next to impossible to manage.

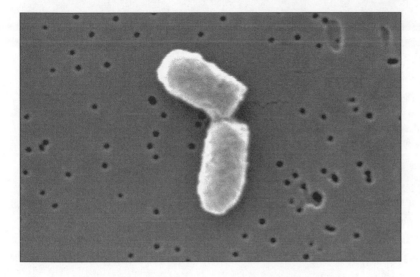

This scanning electron micrograph shows a highly magnified view of Escherichia coli *undergoing cell division, which is how bacteria reproduce. The bacterium is less than a micron wide. Magnification ×21,674.*

(Courtesy of Janice Carr.)

All Together Now

Bacteria often gather in groups: pairs, clusters, or long chains. These groupings tend to be characteristic of particular species and can provide a means of identification.

Bacteria also behave in a coordinated, almost social way, as if they were multicellular organisms. This is especially true of bacteria studied in the wild, rather than laboratory-cultured bacteria. This behavior is called quorum sensing.

In quorum sensing, bacteria communicate with each other by releasing signaling molecules, sometimes called pheromones. A single bacterium can perceive the number of other bacteria around it by measuring the concentration of signaling molecules. When this concentration reaches a critical mass, the bacteria then can adapt to a change in nutrients, carry out defensive maneuvers, avoid toxins, and (in the case of pathogens) even coordinate their virulence so as to evade the host's immune system.

Work on quorum sensing began more than a century ago with the discovery of myxobacteria, a group of bacterial species found in soil and animal dung. Myxobacteria travel in groups, producing large amounts of enzymes that help them digest food. The flocks also produce chemicals that kill off competing microbes, so myxobacteria have become a good source of antibiotics.

E. coli, regarded as a typical bacterium, collects in quorum sensing communities. This scanning electron micrograph shows many rod-shaped E. coli forming colonial groupings, while others have remained isolated as single cells. Magnification ×3,607.

(Courtesy of Janice Carr.)

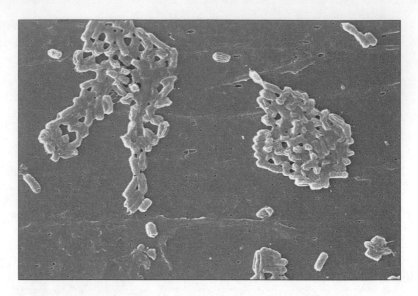

Quorum sensing has now been found in more than 100 species of bacteria. Some researchers suspect it may be nearly universal. Even *E. coli* coordinate growth of their colonies and display pattern formation.

The term *quorum sensing* usually refers to aggregations of a single species, but microbes often form heterogeneous communities of many species as well. *Chlorobium* are members of a group called green sulfur bacteria that photosynthesize, but they do it differently from cyanobacteria. A dozen or so *Chlorobium* species sometimes get together with other bacteria that don't photosynthesize, whereupon they all undergo cell division together. Why? Researchers don't yet know. We'll spend Chapter 9 on microbe social life.

> ### Little Did You Know
>
> The Dutch scientist Anton von Leeuwenhoek (pronounced lay-U-wen-hook), who lived from 1632 to 1723, discovered bacteria in 1683 using a microscope he invented. He did this by scraping plaque off his teeth. Plaque is a biofilm, a kind of microbial community we'll discuss in later chapters, especially in Chapter 9. After he examined his own plaque, he then examined plaque from his wife, his daughter, and from two old men who had never cleaned their teeth. Scientific research is not always pleasant.

Bad Guys, Good Guys

Naturally you know that bacteria cause a lot of diseases: tuberculosis and cat-scratch fever, pneumonia and gonorrhea, ulcers and tooth decay, plague and diarrhea. The list sometimes seems endless. But the fact is most bacteria appear to be either harmless, like the *S. mitis* that went to the moon and back, or helpful, like bacteria that sometimes help give wine its flavor.

We'll spend Chapter 15 on bacterial diseases, but here are a few miscellaneous facts about some fascinating bacteria that *don't* make us sick.

Thermus aquaticus is a hot-springs bug. It makes a unique enzyme that is especially able to withstand heat. So it is part of a process that has become essential to biology: the polymerase chain reaction (PCR), which is used for making unlimited copies of any piece of DNA. PCR is probably the single most important methodological development in the life sciences in the past two decades. That couldn't have happened without *T. aquaticus*, which made PCR affordable and commercially available.

The marine bugs *Vibrio fischeri* and *Vibrio harveyi* are luminescent microbes, but only when they are living in special light organs in particular fish or squid. Dense concentrations of the bacteria switch on a gene for the enzyme luciferase, which causes them to glow. Investigations of this extraordinary bioluminescence have been central to exploring quorum sensing. *V. harveyi* actually has two different quorum-sensing systems. Bioluminescence is thought to attract prey or provide camouflage. The squid *Euprymna scolopes* conceals itself by directing *V. fischeri*'s light to the sea floor and adjusting it to the strength of moonlight, which wipes out the squid's shadow.

Deinococcus radiodurans has an unmatched ability to survive enormous doses of ionizing radiation—a dose six times more than the dose that would kill a human being.

Pseudomonas carboxydovorans can gobble up carbon monoxide and use it as an energy source. This bug is probably the reason that atmospheric carbon monoxide hasn't increased much even though our vehicles have been pumping this deadly gas into the air for decades.

Species of *Lactobacillus* produce cheese, breaking up the sugar in milk to form curds; the curds are then pressed into cheese. Different combinations of microbes produce different kinds of cheese. Holes in Swiss cheese appear courtesy of carbon dioxide produced by *Propionibacterium*.

Yogurt employs *Lactobacillus*, too, in combination with *Streptococcus thermophilus*. These heat-loving bugs turn milk sugar into lactic acid, which thickens the milk and gives it that characteristic tart taste. This partial pre-digestion also means that people who have trouble digesting plain milk can still handle some dairy products. Note that *S. thermophilus* is a good bug. Other kinds of *Streptococcus*, some of them in food (including cheese), can make people very sick.

Bacteria are crucial for beer and wine, too. Alcohol, drinkable and otherwise, is produced initially by yeast, but some bacteria can also convert sugars in plant material to ethanol. This is a popular process in Latin America. The drink pulque is created when *Zymonas* starts digesting the agave plant. Bacteria contribute flavor to alcoholic beverages, but wine- and beer-makers generally consider bacteria The Enemy. *Acetobacter*, for example, turns alcohol into acetic acid, better known as vinegar.

Magnetospirillum magnetotacticum takes iron from the environment and secretes it in cell structures called magnetosomes. Magnetosomes gather in long chains, oriented north to south, and so form a kind of bar magnet inside the bacterium. Since *M. magnetotacticum*'s discovery in 1975, researchers have found other magnetic bacterial species, too. For the moment, they are another ecological mystery. No one knows exactly why having a magnetic personality is useful to a bug.

You may have heard of *Bacillus thuringiensis*, the soil bacterium widely used as a pesticide. It produces a toxin that kills caterpillars and a few other insect species exclusively. The ability to make the toxin has been transferred to a few crop plants with the aim of equipping them to fight off insect pests on their own.

We'll talk in more detail throughout the book about how bacteria live, the illnesses they create, and the jobs they do. For now, what you need to know is that, besides the few Bad Guys, bacteria are crucial to continued life on Planet Earth. Bacteria are the source of about half the oxygen you breathe. Bacteria take nitrogen from the atmosphere and convert it into a kind of nitrogen that plants need for growth. Bacteria produce almost all the antibiotics that cure our diseases, a topic we'll take up later.

Bacteria clean up pollution and gobble up our wastes. And, by breaking down dead organisms and recycling their molecules into new forms of life, bacteria even defeat death.

Archaea, the New Kids on the Block

Archaea were discovered in the 1970s, when researchers realized that genetic material and proteins in some prokaryotes were different from both eukaryotes and bacteria. Researchers believe the ancestor of today's archaea was probably a heat-loving anaerobe that lived before the earth possessed an oxygen atmosphere, 4 billion years ago. Now there are many aerobic archaea species, and oxygen use probably evolved several times. It appears that archaea developed production of the gas methane as a by-product of making energy.

At first scientists called these creatures archaebacteria because their cell structure and metabolism resembled bacteria. But the way they use their genes to make new proteins is more like eukaryotes. So now these prokaryotes are known just as *archaea*, from the Greek word meaning ancient.

Many Archaea Are Extreme Sports, But Not All

Archaea have a reputation as extremophiles, creatures that live in exceptional environments where few organisms can survive. Some archaea species love boiling hot springs and others Antarctic ice. Species are also found deep in superhot ocean vents, or in salty or acid places. Other microbes live in these places too, but, with one or two bacterial exceptions, archaea are the only life-forms that can endure above 100°C (212°F), the temperature at which water boils. These hardiest organisms are believed to survive by making protective molecules that render their cell membranes impermeable.

But archaea also live in more ordinary places, often with other microbes. Researchers have found archaea species in marshes, sewage, soil, and animal intestinal tracts—including yours. Archaea account for an estimated 20 percent of ocean plankton, the tiny organisms that support the ocean food chain.

> **Tiny Tips** _____
> Archaea are different from all other microbes in one hugely important way. So far no archaea species has been found to cause disease in any other organism.

In short, although most known species resemble bacteria, the archaea are quite diverse. They come in many different shapes—among them rounds, rods, spirals, and even polygons—and range in size from 0.1 to 15 μm or more. The group includes both heterotrophs and autotrophs. Some engage in a form of photosynthesis (but not one based on chlorophyll). Their DNA appears to be organized into a ring like bacteria, but their genomes are often much smaller, with few plasmids. Up to half of their genes are unlike any seen in other organisms.

Among the known groups are …

◆ **The methanogens**   Methane-producing anaerobes.

◆ **The halophiles**   Salt-loving aerobes.

◆ **The thermoacidophiles**   Aerobes that live in hot, acid environments.

But there are many other groups as well, and many more remain to be described.

Little is Known About the Archaea

Archaea are difficult to grow in laboratory cultures and therefore hard to study. Scientists do not yet agree about how they should be classified, nor how closely related they are to bacteria and eukaryotes.

Archaea have exchanged genetic material with bacteria during their evolution. Scientists believe that some bacteria borrowed the ability to live at high temperatures from archaea, and archaea borrowed the ability to use oxygen from bacteria. The only way researchers can tell archaea apart from bacteria is to examine their molecules.

The Least You Need to Know

- Bacteria are prokaryotes, live everywhere, are the oldest microbes, and are also the most numerous.

- Bacteria developed photosynthesis, the process that feeds almost all other life-forms and makes the oxygen we breathe.

- Bacteria are classified by several different methods, and there is no single official classification.

- Bacteria usually live in communities of microbes, and sometimes even coordinate their activities.

- Bacteria cause some diseases, but most bacteria are harmless or helpful, making possible life on Earth.

- Archaea, also prokaryotes, may be as old as bacteria but were only discovered in the 1970s. Not much is known about them except that they live everywhere but apparently cause no disease.

Microbes 101, Continued: The Three Eukaryotes

In This Chapter

- ◆ Protozoa, parasites, and more
- ◆ Algae, foundation of the food chain
- ◆ Fungus among us

For 1.5 billion years, the earth belonged to viruses and prokaryotes alone. Then, 2 billion years ago, Earth life took another leap forward with the appearance of eukaryotes.

Eukaryotes, you'll recall, are all those organisms whose genomes are organized into the cell nucleus and surrounded by a membrane. That means animals and plants, of course, but the group also includes many microbes.

Birth of a Eukaryotic Cell

All the microbes in this chapter are eukaryotes. There are three main groups of eukaryotic microbes: protozoa, algae, and fungi.

The eukaryotic cell probably was born when host prokaryotes offered hospitality to other prokaryotes, which moved in and began to specialize in particular jobs. The resulting complex cells were triumphs of mutualism. They could do more things more efficiently. And so they prospered and became the eventual foundation for multicellular life-forms.

Little Did You Know

The eukaryote picture has grown much more complicated recently with the discovery of ultra-small eukaryotes, 20 μm or less in size. Some are as small as bacteria at 0.5 to 2 μm. Some researchers expect these previously unknown life-forms to wreak major changes on existing ideas about eukaryote evolution and ecology. Perhaps they will even alter the definition of eukaryote and shed light on the relationship between the archaea and eukaryotes.

Not so long ago many of these eukaryotic microbes were classified into a separate kingdom, the Protista, often called protists. This group included many one-celled protozoa and algae, but not fungi.

But the Protista were not a natural group of organisms descended from a single common ancestor. Some may have even descended from multicellular organisms. So different protist species were often not closely related at all. Scientists lumped them together because they are not easy to classify as either animal or plant, even though some are a bit like animals and others are sort of similar to plants.

Now science has largely abandoned the idea that these microbes should constitute a separate kingdom. You may still encounter the term *protist* in your readings, however. When you do, just remember that it refers to simple one-celled eukaryotes, either protozoa or algae.

Introducing Protozoa

Protozoa may be just one cell each, but they come in a dazzling array of fantastic shapes. *Protozoa* is the plural form. A single example is called a *protozoan*. Protozoan is derived from the Greek words for "first animal."

Most protozoa are microscopically small, with sizes ranging from 10 μm to 200 μm. More than 60,000 species have been identified, and all need moisture to live. As with all microbes, researchers say many more protozoa are still unknown to science.

Protozoa may be tiny and difficult to classify, but they get around a lot. Like many other microbes, they are everywhere in soil and water and are crucial to the global ecosystem. They consume bacteria, helping to control bacterial numbers and their biomass. Protozoa also eat algae and small fungi, and in turn serve as food for many tiny multicellular creatures.

More than 10,000 species are known to be symbionts and parasites of multicellular organisms, both animal and plant. (You'll recall from Chapter 1 that a symbiont is an organism that lives in or on another organism but does it no harm and sometimes even benefits the host, while a parasite lives on a host while exploiting, and possibly killing, it.) We dedicate Chapter 9 to the topic of symbiosis because it is a major element in microbial life.

The Protozoan Lifestyle

Protozoa are heterotrophs, organisms that live on both organic and inorganic raw materials collected from the environment. (To recap: Unlike autotrophs, heterotrophs don't make their own food. Instead, they eat autotrophs—and each other.) That makes them a bit like animals, because most animals are heterotrophs.

Protozoa abound, especially in polluted waters, and can serve as indicators of environment quality. Terrestrial protozoa like soil and decaying organic matter.

Because protozoa are one celled eukaryotes, each possesses a nucleus. In fact, some of them have two or more nuclei. These organisms don't have much in the way of cell walls, but they are still complex because parts of their one cell are highly specialized.

Specializations include the structures they use for moving around and other organelles called vacuoles. Vacuoles can have some specialized functions in different protozoa species, but their role in all species is to digest nutrients.

A crucial group of specialized cell structures is the mitochondria, the principal source of energy in almost all eukaryotic cells. These organelles are often called the power plants of the cell. You will be hearing much, much more about mitochondria in later chapters.

Most protozoa are free-living, but there is no such thing as a typical protozoan. Protozoa live alone or together. Most protozoa move around, but some are one-celled couch potatoes. Some are unprotected, while others are covered with shells or scales.

Little Did You Know

Heard of the white cliffs of Dover and the pyramids of Egypt? Those limestone structures, both the natural ones and the human ones, are solid foraminifera (or forams), protozoans with external skeletons made of calcium carbonate or sand grains glued together. Forams stay snug in their multichambered shells and dine by sending out strands that form fine netting to catch prey. Nine out of ten forams have become fossils, most laid down in the last 230 million years and turned into limestone. Ancient foram deposits help today's geologists find oil, while modern forams drift down to the ocean floor, forming the white cliffs of the far future.

Like many other one-celled creatures, most protozoans reproduce asexually by binary fission, simple cell division, creating two identical daughter cells. Their chief form of sex is conjugation. Two similar protozoa join together, opening channels between them and exchanging DNA-containing organelles called gametes. Depending on the species, a complicated series of additional cell divisions and exchanges follow, resulting in offspring with a mix of DNA from both original parents.

Like bacteria, some protozoa can enter a dormant stage that protects them in adverse conditions such as lack of moisture. In protozoa, the dormant stage is called a cyst. Protozoa break free of their cysts when favorable conditions return. Cysts can also participate in reproduction and, in parasitic protozoa, transfer between hosts.

Diseases caused by protozoa often occur when cysts transfer from one host to another. That's the way one of the world's most common diseases, amoebic dysentery, is transmitted. People get this serious diarrhea, a.k.a. amebiasis, by unwittingly consuming cysts of the protozoan *Entamoeba histolytica*, often found in contaminated water.

Cysts of Entamoeba histolytica, *the cause of amoebic dysentery, a common form of severe diarrhea. Their hard cell walls permit these protozoa to survive for weeks in the environment until they are consumed by a new host.*

(Courtesy of the Centers for Disease Control and Prevention.)

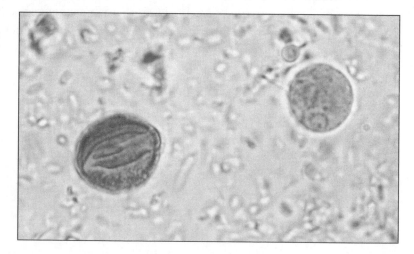

Scientists have in the past classified protozoa by how they move. These traditionally defined groups include …

◆ Flagellates, such as *Euglena*, which possess flagella, whiplike structures that bring in food and help protozoa move.

◆ Amoebas, which shoot out temporary pseudopods, or false feet.

◆ Ciliates, such as *Paramecium*, which possess cilia, hairlike structures similar to flagella, but shorter and more numerous.

◆ Sporozoans, sometimes called Apicomplexa, which are mobile even though they lack flagella, pseudopods, or cilia.

As it turns out, these four groups exist for scientists' convenience, not because the members share evolutionary kinship. Genetic studies have revealed that members of these categories are not necessarily closely related. As is the case with all other microbes, the experts do plenty of arguing about how to classify protozoa and have yet to come to an agreement.

Protozoa and Disease

Ciliates that live in the intestinal tracts of cows and horses are believed to help animals process food, especially difficult-to-digest cellulose, a major constituent of plants. Similarly, a termite couldn't digest wood without the help of flagellated protozoans living in its gut. (Of course humans who own wooden houses don't regard that as a beneficial symbiosis.)

But we know a lot more about harmful protozoa than we do about beneficial ones. Free-living protozoa don't cause a lot of human diseases. Protozoan troublemakers, which have an enormous impact on human health, mostly live as parasites inside other organisms, sometimes in their cells.

There are a great many examples of the harmful protozoa, and we'll examine some of them in Chapter 16. There are protozoa like *Giardia lamblia*, which causes diarrhea, the sexually transmitted *Trichomonas vaginalis*, trypanosomes that cause African sleeping sickness, and many more.

Perhaps the most important are the four protozoa species that live in mosquitoes but cause human malaria. Malaria is one of the world's leading killers, responsible for 1 in 10 children's deaths every year.

While most protozoa that cause human disease live as parasites in other organisms, the reverse is also true. Some protozoa that are otherwise harmless to human health serve as reservoirs of disease because pathogenic bacteria have taken up residence in them. One example is the bacterium that causes the severe respiratory infection Legionnaires' disease. It resides in free-living amoebas that are not themselves pathogenic.

Algae, Not Just Green Slime and Red Tide

Algae live in wet places, too; often they even live in water itself. You're probably getting the picture about microbial names now: *algae* is a plural term. The singular form is *alga*, but it would be a singular event to come across a solitary alga.

Like the other microbes, algae did not descend from one common ancestor, so they are not necessarily closely related and tend to be quite diverse. Treating them as a single group of microbes, as we do here, is largely a matter of tradition and convenience.

Like protozoa, some algae have adapted to extreme environments such as boiling springs and Antarctic cold. Some algae are multicellular and can get very large—for example, seaweeds. Here we'll focus on single-celled algae, most of which range in size from 3 to 10 μm.

Algae Are Plantlike Because Most Depend on Photosynthesis

Algae are eukaryotes and mostly autotrophs. They convert inorganic substances, most notably carbon dioxide, into sugar and free oxygen with the help of light energy. The process is called photosynthesis, which we talked about in Chapter 2. As you'll recall, photosynthesis is the harvesting of light: the conversion of energy from light, most often sunshine, into chemical energy.

def•i•ni•tion

To recap from Chapter 2, photosynthesis uses energy from light, generally sunlight, to build carbohydrates out of carbon dioxide and water. A by-product of this process is free oxygen released into the atmosphere. Photosynthesis is crucial because almost all organisms depend on photosynthesis to provide their food, directly or indirectly. Either the organism is an autotroph that makes its own food, usually with the help of photosynthesis, or it is a heterotroph, an organism that eats autotrophs and other heterotrophs.

Algae carry out photosynthesis in specialized organelles that contain DNA and are known as chloroplasts or plastids. Chloroplasts contain pigments that capture light for the photosynthetic process. The best known of these pigments are the green chlorophylls, which give land plants and green algae their color. But another pigment is also noteworthy: phycoerythrin, a red pigment. Phycoerythrin dominates the chlorophyll in some algae which are, of course, called red algae.

Because most algae use photosynthesis and contain chlorophyll, they have often been regarded as plants, even though they lack structures we associate with plants, such as leaves and flowers. But some algae have lost the ability to photosynthesize altogether. They have become heterotrophs, obtaining all their food from outside sources.

We noted in Chapter 2 that cyanobacteria (incorrectly known as blue-green algae) actually originated photosynthesis long before there were any algae. These bacteria probably contributed photosynthetic machinery to the true algae. The DNA in chloroplasts is very similar to the DNA in cyanobacteria, so scientists believe that chloroplasts originated when cyanobacteria moved into algae and stayed.

Chloroplasts differ in various types of algae, a hint that this symbiosis may have happened more than once. Some scientists disagree, though, holding to the view that oxygenic photosynthesis happened just once. Disagreements of this sort occur all the time in science, and scientific ideas evolve as new information becomes available.

Types of Algae

Some algae live in colonies or are sedentary, but many are free-living amoeboids or flagellates. Green algae are likely ancestors of land plants. Red algae are mostly marine and, with the important exception of phytoplankton, mostly multicellular, which means they are mostly seaweeds; some (such as nori) are popular human food. Brown algae are mostly seaweeds, too. Both red and brown algae are customarily classified as plants, not microbes.

Some algae live inside other organisms and supply a part of their energy through photosynthesis. One notable example are lichens, a symbiosis of a photosynthetic microbe, usually a green alga or cyanobacterium, with a fungus. Together they assume a form completely different from either organism, although sometimes each also can manage on its own. As a result of this mutualism, lichens can live in the Arctic and in deserts, places hospitable to few other organisms. Lichens don't hang out in cities, however, because they don't like air pollution.

Algae engage in both asexual and sexual reproduction. In addition to plain old binary fission, which we learned about in Chapter 2, asexual forms of making new algae include fragmentation and spores. In fragmentation, the alga tissue breaks up and the pieces grow to form new algae. Algae can also produce spores—reproductive structures that are a bit like seeds, but do not contain stored food like seeds. If the spores possess flagella, they can move, but algae also produce nonmotile spores.

Algae Are a Foundation of the Marine Food Chain

Algae are ecologically essential, and particularly significant in watery ecosystems. Microbes free-floating in water, known as plankton, are the first link in most aquatic food chains.

The photosynthesizing plankton, called phytoplankton, are, directly or indirectly, the major source of food for aquatic life (and therefore indirectly the food source for much land-based life as well). Phytoplankton amount to less than 1 percent of the world's photosynthesizers, yet they are responsible for nearly half of the world's biomass (the total weight of all organic matter).

Most phytoplankton are cyanobacteria, but algae are also an important constituent of phytoplankton. These algae usually possess red plastids, not green ones. The majority of algal phytoplankton are either diatoms or dinoflagellates.

Diatoms are said to be the most common algae; a million of them can live in a liter of seawater. Diatoms are as gorgeous as snowflakes, which they resemble a bit. The one-celled diatoms are protected inside lovely shells that are saddled with the unlovely name of frustules. These two-part shells, often compared to two overlapping halves of a petri dish, are an exoskeleton made of silica (that's glass to you).

Diatoms are also enormously important to both fresh- and salt-water ecology and to the economy. Shells from 180 million years' worth of diatoms are mined and used in filters in industry as well as in swimming pools, polishes, paint removers, and fertilizers. Photosynthesis in ocean diatoms generates nearly 20 billion tons of organic carbon every year, and diatoms now have new careers as water quality indicators.

A recent project to learn all about the diatoms' genome makes clear how complicated the evolution of microbes can be. Analysis of the genes of an ocean diatom, *Thalassiosira pseudonana*, reveals that while many of its proteins appear to be related to red algae, others are more similar to proteins in plants and animals. This is just a tease; we'll explore the startling implications of these findings in Chapter 9.

Another algal component of phytoplankton is the dinoflagellates. Have you ever stood on a beach at night and watched the water light up? Chances are that the phosphorescent glow was produced by dinoflagellates. These algae often are round like basketballs and protected with stiff body armor made of cellulose and coated with silica.

The Bad News About Algae

Unlike the other microbes we're discussing, algae don't exactly produce infection and disease. But algae can have serious, and even fatal, health consequences for both wildlife and people. Unfortunately, the lovely diatoms and armored dinoflagellates are also among the microbes responsible for toxic infestations of algae that can deprive other creatures of oxygen and make shellfish unfit to eat.

Algae sometimes grow prodigiously in water, especially when they are oversupplied with nutrients such as excess nitrogen from overfertilized farm fields. These overgrowths—for example, pond scum—are called algal blooms. During a bloom, algae can number millions per milliliter of water.

Little Did You Know

Algae are not the only bad guys in toxic blooms. Cyanobacteria are now known to cause blooms that can prevent oxygen from getting to other marine life-forms and can poison people by contaminating air or drinking water.

Algal blooms are increasing, in part, due to the increase in nutrients pouring into lakes, rivers, and the ocean. Blooms are often unattractive. But worse, they can be toxic to marine life and people. Severe algal blooms in lakes, a process called eutrophication, deprive fish and other wildlife of oxygen. Heavy algal blooms sometimes produce toxins that attack the nervous systems of marine creatures and people.

"Red tides" are usually the result of dinoflagellates, which color the water red when they mass in blooms. Red tides are dangerous because the algae also make a toxin that can paralyze the respiratory systems in vertebrates, including people. Shellfish eat dinoflagellates, but they are not sensitive to this toxin. Unfortunately, people that eat the shellfish are. Paralytic shellfish poisoning is almost never fatal, but it does result in numbness that can last for a few days. A different and more powerful dinoflagellate toxin can build up in fish and, when eaten by humans, can cause diarrhea and respiratory failure.

This satellite image shows a major red tide bloom off Florida's Gulf coast that occurred in December 2001. It extended 30 miles offshore, killed many thousands of fish, and crippled the shellfish industry. The bloom resulted from an overgrowth of the single-celled alga Karenia brevis, *which is harmless at low concentrations.*

(Courtesy of Jacques Descloitres, MODIS Land Rapid Response Team at NASA GSFC.)

The most notorious dinoflagellate at the moment is *Pfiesteria piscicida*, which has been blamed for millions of fish kills on the U.S. East Coast and for making researchers sick with its neurotoxin. This alga is said to have an extraordinary life cycle involving two dozen different stages. The findings have been exceptionally controversial, and at this writing it's not clear exactly how dangerous *P. piscicida* really is.

Diatoms used to be regarded as harmless, but toxic algal blooms have been traced to them recently. Diatom toxins have killed birds and also produce a human disease called amnesic shellfish poisoning, which causes loss of memory.

The Fungus Among Us

The fungi in lichens we mentioned in the previous section are not anything like the fungus you probably know best: mushrooms. Here we go again: *fungus* is singular and *fungi* plural. *Fungus* is the Latin word meaning mushroom. The study of fungi is called mycology, from *mykes*, which is the Greek word meaning mushroom.

But don't be fooled; most fungi don't resemble mushrooms at all. Many fungi are microbes, most notably yeast, mildews, and molds (sometimes spelled *moulds*).

100,000+ Species, but Only a Few Are Harmful

Fewer than 100,000 species of fungus have been identified, but scientists suspect that a million or more kinds of fungi exist in the world. Only 400 of these are known to cause human disease.

Fungi are believed to have arisen about 900 million years ago. For a change, we are dealing with a classification system based on real kinship rather than tradition, because true fungi (as they are sometimes called) do seem to be related to each other.

Many fungal groups appear to have originated on the land, although fungi now live in salt and fresh water, too. Fungi usually prefer warm places, especially the hot, wet tropics, but some live in places where it is cold and even dry.

Fungi are notable for their symbiotic associations with many types of organisms in addition to lichens. Most plants form symbiotic associations with fungi, called mycorrhizae, which help them absorb minerals and water. This relationship began about 400 million years ago. In fact, land plants came into existence because of help they received from fungi. We'll discuss mycorrhizae in Chapter 9. Another group of symbiotic fungi, the endophytes, can affect plant reproduction and sometimes protect plants from herbivores.

Other fungi—rusts, for example—are economically important pathogens of plants. In fact, about 5,000 species of fungi cause most plant diseases. Fungal parasites are responsible for several animal and human diseases, too, from ugly and annoying toenail fungus to fatal pneumocystis pneumonia. Fungi can also produce toxins that make other creatures sick.

Tiny Tips

It's definitely not *microbiology*, but we thought you'd like to know: A humongous fungus in Oregon, measuring 3.5 miles across, is the largest organism on Earth. It's name is *Armillaria ostoyae*, the honey mushroom.

Fungi Are Master Recyclers

In the past, fungi have been thought of as a bit like plants, although recent genetic evidence hints that they are really more closely related to animals.

Their plantlike characteristics are deceptive. They don't do photosynthesis and don't possess chlorophyll, so they are heterotrophs; they must get their food from outside sources of organic compounds. They do this by releasing enzymes that break up large molecules of the organic matter around them into small bits that can be absorbed easily by fungal cells. Because they don't need sunlight to produce food, fungi often live in dark places.

Many fungi make a living by consuming dead matter, breaking it down and making it available for new life. They are often the most important decomposers in an ecosystem. That means they recycle carbon, nitrogen, and other essentials into organic and inorganic compounds that other organisms depend on.

Some fungi literally keep us alive, such as the mold that yielded the world's first antibiotic, penicillin. Also life-giving is the yeast *Saccharomyces cerevisiae*. *S. cerevisiae* helps puts bread, as well as beer, in our mouths. Most fungi are aerobes, but yeasts make a living by changing sugar into carbon dioxide and ethanol (alcohol) in an anaerobic process called fermentation. In baking, the alcohol evaporates while the gaseous carbon dioxide makes dough rise and become bread.

Brewers, however, prize the alcohol; they put it in bottles and cans and sell it as beer. We invented brewing many thousands of years ago. This use of yeast for manufacturing beer is the oldest example of human delight in forcing microbes to make a valuable product. Valuable to humans, that is.

S. cerevisiae is also one of science's favorite experimental organisms. Scientists probably know more about this yeast than any other eukaryote. It is easy to grow in the lab, but since it is a eukaryote, yeast is an excellent stand-in for investigating basic cell properties in complex organisms. For decades it has been essential for the study of genetics. Scientists sequenced its genome in 1996.

Fungal Lifestyles

Yeasts are round or roundish and usually bigger than the average bacterium. They reproduce asexually in two ways: the usual fission or by budding. The buds break away from the parent cell and form new yeasts. Yeasts also reproduce by making spores.

Molds are quite different. That fuzz on the food you forgot about in the back of the refrigerator is made of airy filaments called hyphae. Hyphae are really strung-together cells that grow into a snarl of strands called a mycelium. They often reproduce by splitting up into cells that act like spores.

Some fungi, especially pathogens, can assume two forms: a round, yeastlike one and a tangled mycelium. This shape-shifting is triggered by environmental changes, for example, alterations in available nutrients.

Fungi can have sex, often by fusion of two cells that unites their nuclei. Sometimes this generates spores. Spores are particularly interesting to people who study fungi because their size and shape and colors can help identify species.

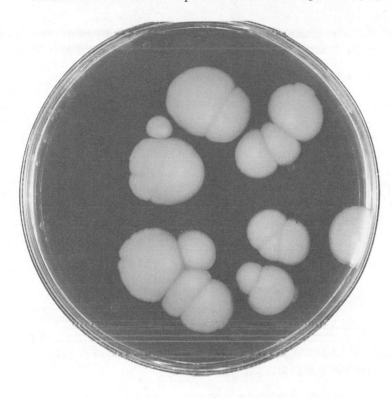

A laboratory culture of the yeast Candida albicans, *a fungus. This organism is normal flora on the human body. But it can sometimes grow wildly, causing infections of the skin and mucosal surfaces known as candidiasis.*

(Courtesy of the Centers for Disease Control and Prevention/ Dr. William Kaplan.)

Alas, fungal spores can also be powerful allergens, as all too many of us know from personal experience. Allergies to molds and fungal diseases of plants have caused more human misery than the human diseases fungi cause directly. The most familiar are athlete's foot and jock itch. Women are cursed with *Candida albicans*, the infuriatingly itchy vaginal "yeast" infection, which can be sexually transmitted. We'll talk more about fungal human ailments in Chapter 16.

Wee Warnings

When is a fungus not a fungus? When it's a water mold, a.k.a. an oomycete. Oomycetes are made of fuzzy hyphae and look like other molds, but they have a different cell structure and scientists don't regard them as true fungi. Water molds are noteworthy because they cause serious plant diseases. At least one water mold changed the course of history: *Phytophthora infestans,* which caused potato blight. In the nineteenth century, potato blight forced millions of starving people to leave Ireland and seek a new life in the Americas.

The Least You Need to Know

♦ Protozoa are a major group of one-celled heterotrophic microbes that contain one or more nuclei and specialized organelles, most notably the eukaryotic cell's powerhouse, the mitochondria.

♦ Protozoa are ecologically important, but they are also an important source of disease for plants and animals, including people.

♦ Algae carry out photosynthesis in specialized organelles, known as chloroplasts or plastids, that contain DNA. They are also a major foundation of the earth's food chain and generate a significant portion of atmospheric oxygen.

♦ Algae don't infect humans and cause disease, but in blooms, they can produce substances toxic to people and animals.

♦ The microbial fungi, which do seem to be mostly related, are numerous, but comparatively few of them cause human disease.

♦ Fungi enter into many symbiotic relationships and are also exceptional at decomposing and recycling organic matter.

Chapter 4

Viruses and Prions: Are They Alive?

In This Chapter

- The parasitical viruses
- Keeping viruses under control
- Phages, the viruses that prey on bacteria
- Cannibals and cattle and prions, oh my!

Up to now we have told you about microorganisms. The prokaryotes and eukaryotes are microbial agents that are living creatures. In this chapter we turn to two kinds of microbes that most scientists believe are not living: viruses and prions.

Viruses, the Skilled Hijackers

A virus is an infectious particle that can grow and reproduce only inside a host cell. That is why viruses are often called obligate intracellular parasites. Like other parasites, viruses infect only a specific group of hosts.

Viruses are no more than dabs of genetic material enclosed in little protein jackets called capsids. A virus usually contains only enough genes to make its capsid and to force its host cell to make more viruses. Some virus capsids, especially viruses that can infect animals, are also surrounded by a membrane called an envelope.

Discovery of Viruses

Viral diseases have been known and treated for thousands of years. Smallpox was even prevented in China and the Middle East by a primitive kind of vaccination. But virology, the study of viruses, formally began toward the end of the nineteenth century, when a number of European scientists investigating disease in the tobacco plant independently concluded that the disease agent must be something other than bacteria. The tobacco mosaic virus (TMV) has since become a favorite study subject for researchers.

Scientists have identified thousands of species of viruses and divided them into over 70 different families, with more recognized every year. They have had a hard time agreeing on how best to classify viruses, so a number of different classification systems exist. Among other systems, viruses have been classified by the following:

Wee Warnings

Are viruses alive? Viruses are neither prokaryotes nor eukaryotes. Because viruses are not truly cells, and in fact must make a living by hijacking the machinery of the cells they infect, scientists argue about whether viruses are really alive or not.

- ◆ The hosts they prefer: animals, plants, insects, fungi, bacteria

- ◆ The type of structure

- ◆ The way they reproduce

- ◆ The kind of genetic material they possess

- ◆ The diseases they cause

Viruses range in size from about 25 nm to about 400 nm, too small to be seen in a conventional microscope. Like other poxviruses, which are the largest animal viruses, the cowpox virus, vaccinia, is a comparative giant, about as big as the smallest bacterium.

Some viruses have tiny genomes, able to make only three or four proteins. Others, like herpesvirus and vaccinia, have much bigger genomes and can make more than 100.

A complete virus particle is called a virion before it enters a cell. A virion is a kind of vector that trundles its genes into the cells of other organisms.

Like other microbes, viruses come in different shapes, but their shapes are not at all similar to those of other microbes. Some viruses, like TMV, are cylinders made of a protein helix. The influenza virus, an enveloped virus, is round and studded with spikes that help it attach to host cells. Some of the spikes contain an enzyme that probably help the flu virus infiltrate its host.

A very common capsid shape is an icosahedron. An icosahedron is a polygon that is similar to a Buckminster Fuller geodesic dome and has 20 different faces. Turns out that's a particularly stable shape when you're building a container or shelter out of a single component—in the case of a virus, a particular protein. Some viruses appear round until they are greatly magnified, but their capsids are icosahedrons.

Viruses have adapted that shape into many forms. Adenoviruses are studded with projections and look a bit like land mines, perhaps not inappropriate, since they cause disease (including colds). Others, notably viruses that infect *E. coli*, look like a vehicle designed for landing on planets: a modified icosahedron with a cylinder extending below and projections that look like legs. Those "legs," called tail fibers, help the virus land on and attach to the bacterium.

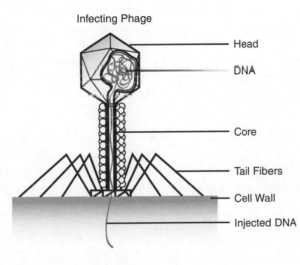

Infecting Phage

Head

DNA

Core

Tail Fibers

Cell Wall

Injected DNA

A schematic drawing of a virus (called a coliphage) that infects the bacterium E. coli, one of the most complex viruses. The virus lands on the bug and attaches to it, and then injects its DNA.

Scientists use viruses to study the cells they infect. They also use viruses as vectors to transport new DNA into host cells. This kind of genetic engineering is covered in Chapter 6. It has many potential applications, including treatment for diseases, commonly called gene therapy. By the way, you are stuffed with viruses. They dwell in many cells in your body, and your DNA is larded with bits and pieces of them. If you have had chickenpox or mono at any time in the past, you have these viruses in some of your cells right now. Scientists estimate that at least 3 percent of the human genome is DNA from inactive viruses. In fact, some of your genes started out as viruses. Prokaryotes have moved in, too—about 0.5 percent of your genes, more than 200 of them, are thought to have come from bacteria.

Where Did Viruses Come From?

There are no fossil viruses, which makes questions about their origins particularly hard to answer. Scientists have a number of ideas about that. Here are two:

- ◆ Viruses used to be independent life-forms, probably prokaryotes, but they moved into other organisms and got lazy. They abandoned most of their previous functions, hanging on only to the genes that permit them to maintain the parasite lifestyle.

- ◆ Viruses used to be functional parts of cells, but they somehow got free, and then, like many a recent college graduate, moved back home.

The question of where viruses came from is not yet settled. It is even possible—some scientists regard it as likely—that viruses arose independently many times in the course of Earth's history.

Viral Diseases

The first human disease to be attributed to a virus was yellow fever, in 1901. Never heard of it? That's because yellow fever, which is transmitted through mosquito bites, is no longer around in the United States, although it's still common in the tropics. Yellow fever is now controlled through vaccination and through controlling mosquitoes.

Wee Warnings

If you learn nothing else from this book, please, please learn this: antibiotics are effective against bacteria, but they do not work against viruses. So next time you get a bad cold, don't nag your doctor for an antibiotic prescription.

Viruses cause more than 400 human ailments, everything from AIDS to cancer, hepatitis to pneumonia, herpes to warts—and don't forget the common cold. But there aren't many cures around (yet) for viral diseases. Our chief weapon against them is prevention by vaccination, deliberately infecting people with a harmless version of a pathogen that cannot cause the disease but that stimulates the immune system to make antibodies against it.

Long before anyone knew anything about viruses, primitive forms of vaccination prevented smallpox in China and Turkey. In the 1790s, the English physician Edward Jenner prevented the disease by inoculating people with pus from cowpox sores.

Smallpox Success Story

Smallpox, caused by the variola virus, is the great success story for vaccination against viral diseases. Smallpox was once a true scourge of humankind that killed one out of three of its victims and often left the survivors covered with disfiguring scars. But the disease has been wiped from the face of the earth with an intensive vaccination program led by the World Health Organization. The last natural case of smallpox occurred in Somalia in 1977.

The virus still exists, locked away in freezers in the United States and Russia. Every so often scientists debate destroying it completely, but they have decided that it's better to hang on to some samples that can be studied anew in case smallpox emerges again or is used as a bioweapon.

Smallpox is notorious for changing the course of history, not once, but many times. Smallpox epidemics in the second and third centuries C.E. probably contributed to the eventual fall of Rome. A Mexico City epidemic that killed as many as a third of its victims also killed the Aztec Empire in the sixteenth century.

Native American populations throughout North America were wiped out by this disease. There is some evidence that on at least one occasion in the mid-eighteenth century, British soldiers tried to infect Indians with smallpox deliberately, one of the earliest cases of biowarfare. Some of today's great Eastern cities, like Boston and Philadelphia, grew up on the ruins of Native American villages laid waste by natural occurrences of disease. They were populated by European colonists who had developed immunity to smallpox because of previous European epidemics.

Another huge vaccination success has been against the great crippler polio. Polio hasn't disappeared completely like smallpox. Although it has dwindled to almost nothing in developed countries, there are still thousands of cases every year in the rest of the world, especially Africa and Asia.

Some Viruses Prey on Bacteria

The viruses that usually interest people are the ones that infect animals and plants, especially the ones that cause disease. Some scientists have been more interested in a different group of viruses: viruses that infect another microbe, the bacterium. They do this by attaching to specific places on the bug called receptor sites, and then pumping in their genetic material.

Viruses that infect bacteria are called bacteriophages (bacteria-eaters) or phages for short. Researchers probably know more about them than any other viral groups. They usually classify phages on the basis of how they look or characteristics of their genetic material. Scientists believe that most of the oceangoing viruses are phages and therefore major contributors to the food web. Phages may also help to control algal blooms.

It is said that for every bacterium, there is a phage. A phage is very fussy about which host it will infect.

There are two kinds of phages: virulent and temperate. Virulent phages get to work the moment they have infected a cell, reproducing speedily. The massive increases in phages cause the host cell to burst open, releasing more phages, a process called lysis.

Temperate phages take it much more slowly. They move their genetic material into the host chromosome or become plasmids, which then get copied along with the host's plasmids. Temperate phages bide their time until the bacterium is about to die, and then launch the process of lysis and releasing new phages.

Sometimes temperate phages even contribute new genetic traits to their host bacterium, a trick that can be bad news for people. One well-known example is a phage that infects a harmless variety of the bacterium *Vibrio cholerae*. Some *V. cholerae* cause the diarrheal disease cholera, a very serious, frequently fatal, ailment. But that happens only when the bug has been infected by a specific phage, one possessing two genes that produce part of the toxin that gives people severe diarrhea.

Phages were the model study subjects of choice for genetics research in the 1930s and 1940s, but became less fashionable as research tools when investigators moved on to eukaryotes. Still, phages have helped scientists study bacteria, and they are also used in molecular biology and as vectors for transferring genes into other cells.

Phages as Weapons Against Disease

Some day phages may turn out to be very useful in fighting bacterial diseases. Late in the nineteenth century, a British researcher named Ernest Hankin noticed that water from some of India's sacred rivers combated cholera. Some years later, the active anti-cholera agents were identified as viruses that infected bacteria.

The cholera-fighting properties of phages led naturally to the idea of using microbes to fight microbial pathogens. Early in the last century, there were reports of phage success against dysentery, typhoid, and plague. Bacteriophage therapy had a brief heyday, especially in the 1920s. But results on other diseases were mixed, and with the appearance of antibiotics in the 1940s, phage therapy faded away in the United States, although it has continued in Russia and Eastern Europe.

Perhaps medical history is about to come full circle. Phage cures were vanquished by antibiotics, but now the burgeoning problem of bacteria that are resistant to antibiotics has led to new research interest in phage therapy.

Phages do kill pathogenic bacteria effectively, and they do it without penetrating human cells, which they can't even recognize. They accomplish this through lysis.

Technical problems are delaying the appearance of phage therapy in the clinic, but one solution may be an approach that does not rely on phages themselves. Instead it employs enzymes that phages produce to smash their way out of their host bacteria so they can infect new hosts.

Researchers are experimenting with employing these enzymes externally (in nasal sprays, for example) to kill bacteria. So far they have achieved success with several kinds of *Streptococcus* as well as the bug that causes anthrax and a few other bacterial pathogens. The beauty of this system is that the enzymes only kill the pathogen, not, like antibiotics, all the surrounding harmless (or helpful) bacteria.

Some Viruses Prey on Eukaryotes

Viruses that infect animal cells can harm those cells in many ways. These viruses can …

- ◆ Produce viral proteins that have a direct toxic effect on host cells.

- ◆ Interfere with the work of an animal's genetic material and its protein manufacture.

- ◆ Create new cell structures, known as inclusion bodies, to demolish the host cell.

- ◆ Smash host cell structures that contain digestive enzymes, releasing the enzymes to damage the cell.

- ◆ Alter the surface of the host cell so that it will be attacked by the host's own immune system.

- ◆ Alter the host cell's growth cycle, turning it into a cancer cell.

We talk in detail about the diseases these animal infections cause in Chapter 14.

Although more research has been focused on animal viruses, plants also suffer from viral diseases. These diseases are often carried from plant to plant by insects. Once inside a plant cell, the virus can travel to other parts of the plant through its circulatory system. Cells infected with TMV, the best-studied plant virus, can develop inclusions, usually consisting of clumps of virions. Plant viruses can also disrupt the chloroplasts that are essential for photosynthesis.

There are also viruses that infect algae and fungi, but not much is known about them.

Little Did You Know

Several plant diseases are caused by odd entities that are even simpler than viruses. Known as viroids, they are essentially naked pieces of genetic material. This genetic material is not DNA, but rather RNA, which is discussed in Chapter 6. Viroids "reproduce" by inserting themselves into the host plant's nucleus, where they are copied along with the host's genome. Viroids usually spread in the host plant's seed.

Scientists have grown increasingly interested in insect viruses because of their potential for controlling insect pests without the use of toxic chemicals. Attention has focused in particular on viruses that infect insects only.

Insect viruses also produce virion-stuffed inclusion bodies in host cells, and one potential method for fighting insects is to spray these inclusion bodies, with their deadly cargo of virions, onto plants that insects eat. Another approach is to release infected insects in hopes they will pass the infection along. As you may imagine, these tactics are controversial because of uncertainty about the long-term effects of deliberately spreading insect viruses in the environment.

Prions, Cannibals, and Mad Cows

The story of human prion diseases began with the grisly tale of cannibalism on the island of New Guinea. Members of New Guinea's Fore tribe suffered from a devastating brain disease called kuru. In 1961, researchers decided the disease might be due to one of the tribe's burial rituals. It was Fore custom to remove, handle, and sometimes eat the brains of dead relatives.

The investigators theorized that the brains of people who died of kuru contained the organism that caused it, and the tribe's funeral rite helped the deadly organism transfer to new victims. When tribe members stopped the custom, kuru vanished.

A few years later, researchers demonstrated that lab animals developed kuru after being injected with brain tissue from someone who had died of the disease. This confirmed their hypothesis that kuru was an infectious disease.

Although they were unable to find a disease-causing organism, researchers believed it was probably caused by a virus. Because kuru symptoms take a long time to appear after infection, they called it a slow virus. Since then, several nervous system diseases have been attributed to slow viruses, but no one has ever found such a virus.

Most people, however, have heard of prion disease because of mad cow disease, also known as bovine spongiform encephalopathy (BSE; *encephalopathy* is a Greek-derived word for sick brain). This horrible disease emerged in Britain in the 1980s. It has resulted in the deaths of thousands of British cattle and also has killed hundreds of people who ate beef from infected animals.

British cattle are believed to have contracted mad cow disease by eating animal feed that contained protein meal made from sheep carcasses infected with scrapie, an encephalopathy found in sheep and goats. Similar diseases also have been found in other animals, including cats, mink, and elk, and the chronic wasting disease that afflicts white-tailed deer.

Prions are not living organisms; they are proteins that can cause disease. At first, most scientists laughed at the idea that infection could be caused by an entity that has no genes and is not alive, but the evidence has grown. A few holdouts still argue that a cofactor, probably a virus, causes prion diseases, but most scientists have come to accept that prions, while not alive, can still be agents of disease—deadly disease.

In 1982, researchers suggested that both kuru and scrapie were due to abnormal brain proteins. Stanley Prusiner named them prions, short for proteinaceous infectious particle. In 1997, Prusiner's work earned him the Nobel Prize in physiology or medicine.

Prions Fold Other Proteins into Abnormal Shapes

A prion is an abnormal version of a protein called PrP^c that usually sits on the surface of brain cells. No one yet knows what the normal protein does, although it is similar to a protein involved in Alzheimer's disease. Prions have the same structure as PrP^cs, but have been twisted out of shape, which changes their properties.

When a prion comes in contact with PrP^c, it transforms the normal protein into a misfolded prion (called PrP^{sc}). The brain breaks down the normal protein all the time, recycling its parts into other chemicals. But because of their different 3-D structure, prions clump together and resist being taken apart and recycled. They also withstand radiation, chemicals, and even heat that would be deadly to other disease agents. As they accumulate, they destroy brain cells.

Whether they attack cows or people, the prion diseases are all similar and all gruesome. Prions kill neurons, chewing holes in the brain, which leaves it looking like a round, whitish sponge. This is why prion diseases are called transmissible *spongiform* encephalopathies (TSE).

> ### Little Did You Know
>
> The best-known human TSE is Creutzfeldt-Jakob disease (CJD), which in most cases appears for no obvious reason. It's not a common disease; fewer than 1,000 cases total are known in the United States. The human TSE acquired through eating beef from animals with mad cow disease is called variant CJD (vCJD).

Prions also lie dormant for many years, causing no problem, then suddenly trigger dementia and other symptoms such as trembling and difficulties with movement. The diseases are almost always fatal within a year after symptoms begin. Researchers are trying to develop treatments, but they're not there yet.

Prions are believed to cause other human dementias besides kuru, all of them rare. Strangely enough, sometimes these diseases are inherited via a gene that makes abnormal PrP^{c}s, but sometimes they are transmitted by infection—for example, eating prions (as in kuru), transplanting infected organs, or even improperly sterilizing surgical instruments are ways these diseases can be spread.

The Least You Need to Know

- Viruses can't survive on their own, but reside in other cells and hijack a cell's genetic machinery.

- Virus diseases cannot be cured with antibiotics like bacterial diseases, but they can be prevented by vaccination.

- Viruses are fussy about their hosts, so there are viruses specialized for infecting animals, plants, and even other microbes.

- Phages are well-studied viruses that infect bacteria and have the potential for fighting many diseases.

- Prions are definitely not alive, but they cause strange, deadly brain diseases in animals and people by converting normal proteins into abnormal forms.

Part 2

How Microbes Make a Living

This section is just what it sounds like. It discusses how microbes live. Here we give a quick rundown on the chemistry of life, a little about microbial genetics and genomics, cells and cell structure, and how microbes eat and grow and (blush) have sex (although most of them don't, sorry to disappoint you).

5

The Molecules of Life

In This Chapter

- ◆ A bit of biochemistry
- ◆ Chemical bonds: how molecules stick together
- ◆ A brief immersion in water
- ◆ Keeping things organic
- ◆ DNA and RNA

Microbes are so much tinier than anything else in our everyday experience that it's almost impossible to think about their size. And of course it *is* impossible to see them at all without the help of powerful magnifying lenses. Yet microbes, and everything else, are made of matter, and matter is made of things tinier still: elements. Elements are made of even tinier things.

What these things are and how they work together in life—in short, biochemistry—is the subject of this chapter.

A Brief Romp Through the Chemistry of Life

Elements are substances that cannot be broken down into simpler substances using chemical methods. There are more than 100 elements, but only 92 occur in nature; the others were created in science labs. Among the natural 92 are element names that will probably be familiar to you, such as hydrogen, carbon, nitrogen, and oxygen. These are among the most common elements, which is a good thing, because they are essential for life.

But wait, things get smaller still. Elements are made of molecules, which are the tiniest chunk of an element that still retains all the chemical properties of that element.

Molecules are made of something smaller as well: at least two atoms. Atoms are the building blocks of elements. Guess what? Atoms, of course, are made of particles, too. A typical atom consists of three kinds of particles:

- Protons, which have a positive electrical charge

- Electrons, which have a negative electrical charge

- Neutrons, which have no electrical charge

You can visualize an atom as looking a bit like an incredibly tiny solar system. It's made of a dense central core composed of protons and neutrons surrounded by a cloud of orbiting electrons. The number of each kind of particle depends on the element. Hydrogen, the smallest atom, is made of just one proton and one electron; there are no neutrons at all. Carbon has six each of protons, neutrons, and electrons.

By now it will not surprise you to learn that subatomic particles like protons are made up of even smaller particles: for example, quarks. But quarks and company are the province of physics, not microbiology, so that is the last you'll hear about them here. We talk some about atoms in this chapter. But the building block of matter that is most important to microbiology is the molecule. A molecule consists of at least two atoms, and usually many more.

Linking It All Together with Chemical Bonds

How do atoms stick together to form molecules? Atoms "prefer" stable configurations, where the number of protons in the nucleus—the dense, positively charged center of an atom—equals the number of electrons orbiting the nucleus. An atom is most stable when its outer orbital shell is filled with the right number of electrons. So it seeks to maintain stability, either by snagging missing electrons or getting rid of extra electrons.

Atoms add or lose electrons with the help of chemical bonds. There are several kinds of chemical bonds. Here are three you should know about because they are particularly important in the molecules of life:

- ◆ Covalent bonds, where atoms join together because they share pairs of electrons

- ◆ Ionic bonds, where atoms are attracted to each other because they possess opposite electrical charges

- ◆ Hydrogen bonds, where hydrogen atoms and atoms with a negative electrical charge bond, often with oxygen or nitrogen

These bonds almost always form between atoms that are already part of other molecules, not single atoms. Covalent bonds are usually the strongest of the three. Ionic bonds are often a bit weaker, and hydrogen bonds weaker still. But even weak bonds can bind the parts of a molecule together quite tightly if there are a great many of them.

Getting Our Feet Wet

Life on Earth is based on water, which is why scientists hoping for life elsewhere in the universe are scouring planets for signs that they ever possessed water. Eukaryotes are some 70 percent or more water by weight. Some microbes have adapted to near-dry conditions, but they must use extra energy to hang on to what water they have. Most microbes can't grow at all without plenty of water. That's one reason drying food often preserves it from spoiling.

Water is a simple molecule, two atoms of hydrogen, each sharing its one electron covalently with an atom of oxygen. You know the formula of course: H_2O.

Overall a water molecule is electrically neutral. But the oxygen atom is more electronegative, and the hydrogen atoms are more electropositive. So the water molecule is shaped like an isosceles triangle, or perhaps Mickey Mouse, with a body and two big ears. It is a bit more negative on the oxygen (body) side and a bit more positive on the two hydrogen (ear) sides. That means it can easily form hydrogen bonds with other water molecules, joining its negative side to another's positive side and vice versa.

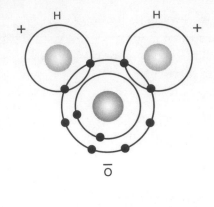

Water is an example of covalent bonding, in which atoms join together by sharing electrons. You can visualize the atoms in a water molecule by visualizing a silhouette of Mickey Mouse. An atom of oxygen, surrounded by its electrons (small dots), shares electrons with two hydrogen atoms. The molecule can also form hydrogen bonds with other water molecules.

At temperatures below freezing, these hydrogen bonds are pretty stable, and water becomes ice. At higher temperatures, in liquid water, the hydrogen bonds break and form constantly, a trillion times every second.

You may have heard water referred to as the universal solvent. It's not quite true, of course, that water can dissolve anything, but it can dissolve many things. That property is a critical component of the chemistry of life.

Here's an example of this process using ordinary table salt, composed of one atom each of the elements sodium (Na) and chlorine (Cl). The formula is NaCl. Salt dissolves in water because the positive sodium and negative chloride separate into their individual ions and tend to stay that way, each one encircled by water molecules that keep it from linking to an ion with the opposite charge.

The Organic Molecules in All of Us

Nothing very complicated here: organic molecules are molecules that contain the ubiquitous element carbon. As you probably know, we are carbon-based life forms—but so is every other organism. Carbon is essential to life, or at least life as we know it.

Carbon "wants" its outer shell to possess eight electrons. It usually has only four, so carbon tends to form four covalent bonds. It hooks up with many other elements: oxygen, for example, to form the carbon dioxide (formula CO_2) essential to plant life. Carbon bonds frequently to nitrogen, phosphorous, and sulfur as well. Carbon also bonds easily and firmly to itself, forming very stable long chains and rings.

There are many kinds of organic molecules. In this chapter we talk about the following:

- Carbohydrates

- Lipids

- Proteins

- Nucleic acids

Wee Warnings

Not all compounds that contain carbon are considered to be organic. For example, carbon dioxide is not an organic compound. At a minimum, organic compounds contain both carbon and hydrogen that are covalently bonded.

Sweet Talk About Carbohydrates

That's carbs to you. Carbohydrates are one of the three chief means of storing and using biological energy. The other two are lipids and proteins, which we get to in a minute. Carbs include sugars and starches.

Carbs consist of carbon chains with hydrogen and oxygen attached, often in a ratio of 1:2:1; that is, one carbon to two hydrogens and one oxygen. Carbohydrate molecules can get enormously big and complex, but basically they consist of small, mostly repetitive modular units linked together in diverse ways. The basic unit in carbohydrates is one carbon combined with one water, i.e. carbo hydrate.

The simplest carbs are called monosaccharides. One example is the simple sugar glucose, the fuel your body runs on. When monosaccharide units link together, the resulting carbs are called disaccharides (two monosaccharides), oligosaccharides (several monosaccharides, usually between three and nine), or polysaccharides (many monosaccarides, usually more than ten).

An example of disaccharides is sucrose, ordinary table sugar, which is made from one molecule of glucose and one of another monosaccharide, fructose, a sugar commonly found in fruit. Few oligosaccharides exist naturally; most are created when polysaccharides break down.

Polysaccharides such as starches are often called complex carbohydrates. The cellulose that makes plants rigid is a complex carb. Polysaccharides can be very complex indeed, sometimes branched constructions made of thousands of sugars. Disease-causing bacteria often produce a thick, slimy layer of polysaccharide, using it to conceal themselves from a body's immune system. If the immune system is prevented from recognizing the bacteria, it can't launch its defensive measures. The common gut bacterium *E. coli* is able to make some 200 polysaccharides.

Lipids: Fat City

Lipids, better known as just plain fats, are another way of storing and using energy. But lipids have other biological functions, too. For example, they serve as building blocks for cell structures. By contrast with the relatively simple, mostly modular structure of carbohydrates, lipids come in a dizzying array of forms.

As a group, these organic molecules resist dissolving in water, a notable exception to the notion of water as a universal solvent. Water is weakly charged, and so is every-thing that dissolves in it. But lipids have no charge, thus they cannot dissolve. That makes lipids quite good at keeping an organism from losing water, so they are impor-tant constituents of cell membranes.

That's where the often-asymmetrical structure of lipids comes in handy. One example is phosphatidyl ethanolamine, found in bacterial membranes. This molecule has a lipid end with no electrical charge so that it is water-repelling (hydrophobic). This end is lodged deep inside the membrane, forming a protective barrier guarding the bacterium's water supply. The other end does have a charge and is attracted to water (hydrophilic). It sits at the membrane surface.

The Big and Important Proteins

The third type of energy-storing molecules are proteins. But that ain't all. Proteins are big, complicated 3-D organic molecules that do big, complicated things in organisms. They are essential for the structure, function, and regulation of all cells and tissues in all living things.

Each protein has a specific job. Proteins serve as enzymes or parts of enzymes that smooth the progress of chemical reactions in organisms; some enzymes consist of only one protein, while others function as a multimeric unit. Proteins are the molecules of the immune system that help ward off disease. They form the structural supports of cells, which we discuss in Chapter 7. They handle signaling and communication within and between cells. They are food. Some proteins carry out multiple functions.

Besides carbon, proteins contain the elements nitrogen, hydrogen, oxygen, and often sulfur. Proteins are made of chains of amino acids, folded into 3-D structures.

Stay alert; a heavy load of definitions is on the way. An amino acid is a molecule that contains two specific kinds of functional groups (a set of atoms that governs a sub-stance's chemical behavior). One of those essential functional groups is an amino group: a nitrogen atom covalently bonded to two hydrogen atoms, formula NH_2.

The other required functional group is carboxylic acid, a molecule containing a carboxyl group. And what is a carboxyl group? An atom of carbon with a covalent double bond to an oxygen atom and a covalent single bond to a molecule combining one atom each of hydrogen and oxygen, known as a hydroxyl group. That sounds more complicated than it is; a carboxyl group is just four atoms, and the formula is COOH. COOOOOL!

There is just one more definition of amino acid. In biochemistry (which this is, in case you hadn't caught on yet), amino acid usually means an alpha amino acid. This is an amino acid in which the amino group and the carboxylic acid are attached to the same atom of carbon.

Proteins are held together by a kind of chemical bond we haven't discussed yet: the peptide bond. A peptide bond results when the amino group (NH_2) of one molecule reacts with the carboxyl group (COOH) of another. That reaction throws off a molecule of water (H_2O), leaving behind a bond: C-NO. It's called a peptide bond because when amino acids are linked together in particular orders they are called peptides. You can think of peptides as small proteins or pieces of proteins, each one with its characteristic amino acid sequence.

> **Little Did You Know**
>
> Only about 20 different kinds of amino acids are normally found in proteins, although more than 100 exist in nature. Some microbes can make unusual amino acids. Humans, unlike some microbes, cannot synthesize all 20 different kinds of amino acids required to sustain life. That is why it is important that our diets contain the missing amino acids (nine of them!).

You'll recall that we said that proteins fold into complex 3-D structures. Scientists talk about those structures in four different ways:

- Primary structure, the sequence of amino acids that make up a particular protein.

- Secondary structure, the general shape of the molecule or of particular parts of the molecule, a consequence of the way a polypeptide is organized in space around a single axis.

- Tertiary structure, how the secondary structures of a single molecule of protein are organized around three axes to become a 3-D shape.

- Quaternary structure, the more complex 3-D shape that results when two or more *protein subunits*, which are polypeptide molecules with their own primary, secondary, and tertiary structures, get together to form part of a larger chunk of protein (which itself may be only part of the entire protein).

To further complicate the protein story, these 3-D structures change as proteins do their work. When these changes involve tertiary or quaternary structures, they are called *conformations*, and the switch from one state to another is called a *conformational change*.

When we said that proteins are held together by peptide bonds, we were talking about what binds the primary structure of proteins, which are covalent peptide bonds. Hydrogen bonds link secondary structures and figure in tertiary structures, too, but so do other kinds of bonds.

Primary structure determines how the other structures form. This process is called *protein folding*. Scientists haven't quite figured out everything about how protein folding works yet, so it is the subject of intensive research. That's because protein folding is central to understanding the behavior of proteins, which is flexible and changeable depending on their conformation.

Polypeptides can fold in different ways. Some of those conformations will be abnormal and prevent the protein from working correctly, but for each polypeptide at least one conformation is stable, often more than one. That's crucial. Why? Because a different folding pattern will usually cause a polypeptide to work in a different way.

That means that two proteins with identical primary structures, that is, identical amino acid sequences, can have completely different biological functions depending on which way they are folded. This flexibility, where the same molecule can do different things depending on its shape, is a key to understanding why there are so many weird and wonderful life forms, so don't forget it. But it can also make problems. Remember the prion story, where misfolded proteins bore holes in the brain that lead inevitably to a horrible death?

Nucleic Acids: DNA and RNA

Nucleic acids are another type of complicated molecule. Their main job is to store and transfer the genetic information that governs the behavior of a cell.

Thanks to television crime shows, you've probably heard a lot about one nucleic acid: DNA. You may have also heard of another one: RNA. These are the genetic molecules that issue instructions for building cells and bodies. A big part of their job is to give orders for making all those proteins we've been talking about. In addition to its roles in a cell's genome, RNA can also act as an enzyme and is a required component in some enzyme complexes.

In this chapter, we discuss how DNA and RNA are put together, which will help you understand how they do what they do. We talk about what they do in Chapter 6.

Despite their size, DNA (*deoxyribonucleic acid*) and RNA (*ribonucleic acid*) do not have a terribly complex structure, nothing like the complexity of most proteins. The nucleic acids are simply very long chains of a handful of standard modular subunits called *nucleotides*. The order in which the thousands of nucleotides are strung together encodes genetic information, just as the order of letters in sentences encodes information.

A nucleotide is a five-carbon sugar bonded covalently to one of five molecules called bases. The bases are made of atoms of carbon (of course), hydrogen, oxygen, and nitrogen in various combinations. The bases are named adenine, thymine, guanine, cytosine, and uracil. Each base is joined to the next one by a phosphate group (a phosphorous atom plus four oxygens: PO_4). Usually the bases are abbreviated A, T, G, C, and U. The bases A, C, and G are found in both DNA and RNA, whereas T is only in DNA and U is only in RNA.

DNA is an enormous molecule, almost always formed from two chains of nucleotides wound together in the famous double helix. (One exception: some viruses use single-stranded DNA.) Double-stranded DNA looks a bit like a very long spiral staircase or a twisted ladder. The two twisty sides of the helix (or ladder) are formed from sugars and phosphates, often referred to as the backbones of DNA.

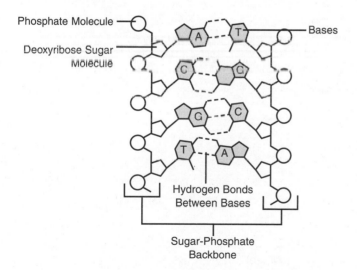

A bird's-eye view of one base pair in the DNA double helix. The complementary bases are linked together by hydrogen bonds (dotted lines).

The "treads" of the staircase (or rungs of the ladder) are stacked on top of each other in the helix and formed from pairs of bases. Adenine in one DNA strand always pairs with thymine in the other, and guanine always pairs with cytosine. In RNA, uracil replaces thymine and pairs with adenine. The bases are joined in the middle of the rung by hydrogen bonds. As you'll recall, hydrogen bonds are not particularly strong, so it takes only a small amount of energy to split the rungs in half and separate the two strands of DNA in the helix.

The firm rule about base pairing, A with T and G with C, means that the two individual strands of the DNA double helix are *complementary*. The strands can be unzipped from each other, and each one can go off on its own. But it can only pair up again with a complementary strand, one that duplicates the base sequence of its previous partner. You can see that this pairing rule is a big help in preserving the structure of a particular stretch of DNA, ensuring that the molecule gets copied accurately and its genetic information is conveyed correctly.

DNA makes new copies of itself by unzipping. Breaking the hydrogen bonds in the middle of the "ladder," it creates two separate strands. Each strand then forms a new partner strand by seeking its complementary bases. The result: two new DNA helices, each a duplicate of the original.

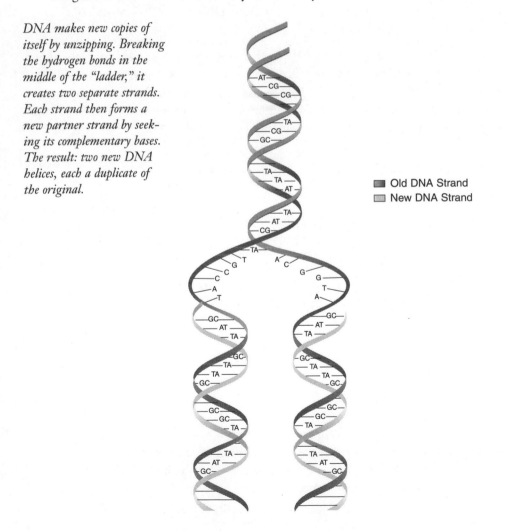

■ Old DNA Strand
■ New DNA Strand

Although the nucleic acids are an essential part of every living creature, they behave differently in prokaryotes and eukaryotes. In eukaryotes, much of a cell's DNA resides in the nucleus. Prokaryotes, you'll recall, do not possess a nucleus. Instead their DNA molecule is organized into a circle.

RNA is not the same chemically as DNA. Its sugar is a bit different, and so is its base pairing: adenine pairs with uracil rather than thymine. RNA is also typically just a single strand, not double-stranded like DNA. However, RNA can coil back upon itself so sections of the single strand can form a double helix. The two molecules have quite different functions, as we talk about in Chapter 6.

Single-stranded RNA, showing its bases. RNA does not use thymine; adenine (A) pairs with uracil (U) instead.

Ribonucleic Acid (RNA)

The Least You Need to Know

◆ All living things, including microbes, are made of chemical elements, which are substances that cannot be broken down into simpler substances using chemical methods.

◆ Atoms are the building blocks of molecules and are made of three kinds of particles: protons, which carry a positive electrical charge; electrons, which have a negative electrical charge; and neutrons, which have no electrical charge.

◆ Organic molecules are molecules that contain the elements carbon and hydrogen, and several of them are essential for life.

◆ Carbohydrates are organic molecules like sugar and starch, used for storing and using energy and consisting of small, mostly repetitive, modular molecules linked together in diverse ways.

◆ Proteins are big, complicated 3-D organic molecules made of chains of amino acids that are essential for the structure, function, and regulation of all cells and tissues in all living things.

◆ The nucleic acids DNA and RNA are another type of complicated molecule whose main jobs are to store and transfer the genetic information that governs the behavior of a cell.

Chapter 6

A Little About Microbial Genetics and Genomics

In This Chapter

- ◆ What genes do and how they do it
- ◆ Transcription and translation
- ◆ Transposons, the jumping genes
- ◆ Gene sequencing
- ◆ Microbial genome projects
- ◆ Genetic engineering and microbes

All life-forms need genes to tell their cells what to do, and microbes are no exception. Even viruses, which are not life-forms, have genes; in fact, viruses are nearly all genes.

In the last chapter we told you something about the structure of nucleic acids, the genetic material DNA and RNA. In this chapter, we present genetic basics, explaining how DNA and RNA do what they do.

Genetic Basics

You and the microbes may seem utterly different, but you are related. All the earth's billions of life-forms are kin to each other.

How did we and our very distant cousins come to look so different and develop so many different ways of getting on in the world? A century ago we began to answer that question with the help of a new science called genetics.

Genetics investigates how the genetic material in cells affects what goes on inside those cells. Chemical reactions within an organism's cells create its physical characteristics. These reactions are governed in part by genes and other genetic material and in part by things that are not genetic material. Scientists have only begun to grasp the near-unimaginable intricacy of the complex dance of genes and environment that results in a hot springs life-form or a fungus—or you.

What Are Genes?

Genes encode proteins that monitor and control all the activities in cells. Genes are made up of specific strings of genetic material. *Genomics* is the study of the full set of an organism's genetic material.

The best-known genetic material affecting what goes on inside cells is deoxyribonucleic acid, DNA for short. DNA is just information, rather like a vast library. A gene is a bit like one book in that library. Genes are part of long strings of DNA called chromosomes. You can think of a chromosome as a single bookcase in the big library.

But these genetic library books are written in a coded alphabet. The code contains a set of instructions that tells cells what to do. You'll recall that this DNA code is an alphabet of just four chemical "letters" known as nucleotides or bases. The four nucleotides are named adenine, guanine, thymine, and cytosine. For short, scientists call them A, G, T, and C, respectively.

Tiny Tips

DNA is the universal genetic code, but some viruses use a slightly different but related code: ribonucleic acid, known as RNA. We talked a bit about RNA in Chapter 5, and we talk more about it later in this chapter.

Even though there are just four of them, the nucleotides can be strung together in billions of ways. That means DNA can send billions of different coded instructions to cells. And if billions of instructions are possible, we can begin to grasp how the archaea, bacteria, viruses, algae, and fungi can be so different from each other, and so different from fish and frogs and flamingos and foxes, and people, even though all these organisms are related.

People have, of course, long known that living things inherit traits from their parents. Our long-distant ancestors made that common-sense observation 10,000 or more years ago. That's about when they invented agriculture, which is the purposeful breeding and cultivation of plants and animals for desirable characteristics. But exactly how traits were passed to the next generation was a mystery until the late 1800s. Now we know that the same general principles govern the growth and development of all life on Earth.

What Do Genes Do?

Genes do their work by influencing what goes on inside cells. How do they exert that influence? Structural genes issue instructions for making proteins. Proteins are big, complex molecules that carry out most of the work involved in keeping an organism going. See Chapter 5 for a refresher.

Structural genes, the ones that code for proteins, are what scientists mean when they use the word *genes*. Structural genes are the genes scientists have studied the longest and know the most about. But there's more to DNA than structural genes.

Thanks to proteins, cells and the organisms they form arise, develop, live their lives, change, and create descendants. They do all these things with proteins made according to the instructions in their genes. The way those instructions are carried out is shaped by the world around them.

Proteins do two essential jobs in the cell. First, they are its main building materials, forming the cell's architecture and structural components. Second, proteins do most of a cell's work.

In addition to proteins forming a cell's bricks and mortar and fixtures, proteins also compose …

- ◆ Antibodies that identify invading organisms and foreign proteins.
- ◆ Transport molecules that carry oxygen to where it's needed.
- ◆ Enzymes that carry out chemical reactions inside the cell.

Thousands of chemical reactions take place in a cell. Enzymes use a process called catalysis to get the reactions going. Enzymes help make other molecules, including DNA. They also break down food and deliver and consume the energy that powers the cell.

Other kinds of proteins, called regulatory proteins, preside over the many interactions that determine how and when genes do their work and are copied. They also supervise enzymes and the give-and-take between cells and their environment.

A small part of the genome is devoted to regulatory sequences. These are chunks of DNA whose function is gene regulation. They govern gene expression, such as defining the beginning or end of structural genes and turning those genes on and off.

Even though these regulatory sequences are DNA, too, they are not usually called genes. This can be a bit confusing. When some scientists use the word *gene*, they mean a structural gene plus its regulatory sequences, not just the structural gene by itself.

How Do Genes Work?

To carry out its many functions, a cell constantly needs new copies of proteins. Although proteins can do lots of jobs well, they cannot make copies of themselves. To make more proteins, cells use the manufacturing instructions coded in their genetic material.

It's a stupendous assignment. As you'll recall from Chapter 5, proteins are huge, complex molecules that must be folded into intricate 3-D shapes in order to work correctly. They are made out of various combinations of 20 different chemicals called amino acids. The DNA code of a gene—the sequence of its "letters" A, T, G, C—spells out the precise order in which the correct amino acids must be strung together in order to form particular proteins.

Sometimes there is a mistake in those instructions, a kind of typographical error in the gene book. The mistakes are called mutations. A mutation is simply a change in the DNA sequence. The word *mutation* is commonly used to mean a harmful alteration in DNA. But its true meaning is neutral, so let's repeat: a mutation is simply a change, any change, in the DNA sequence.

That doesn't mean the change is always unimportant. Type "lamp" when you mean "lamb" and you change the meaning dramatically. These are both objects, but very different objects.

DNA has several ways of repairing mutations and usually does a wonderful job of fixing the mistake. But if they are not repaired, mutations can cause the genes to work incorrectly or even not at all. The result is an abnormal protein, or perhaps no protein.

Not all mutations are harmful. Many have no particular effect on an organism's life. Sometimes a mutation assists an organism to do better, and it passes on that mutation to its descendents. If a mutation helps the organisms leave more descendents, eventually the mutation will be permanent and become a normal part of the DNA of that group of organisms. In fact, mutations are the engine that drives the evolution of new life-forms.

So How Does DNA Make Proteins?

DNA doesn't make proteins. DNA is just a collection of instruction manuals. The instructions are transmitted to a cell's protein factories by RNA.

Tiny Tips

RNA is not nearly as well known as DNA, but it is as remarkable. In fact, many scientists believe that RNA appeared on Earth long before DNA and is actually its ancestor. That's why some viruses that arose very early in Earth's history, before DNA came into existence, operate with RNA instead of DNA. Among these are serious pathogens: HIV (the AIDS virus), Ebola virus, and viruses that cause polio, SARS, and West Nile fever, among others.

Chemically, RNA is very much like its offspring DNA, with one significant difference in its genetic code. The RNA code uses the bases adenine (A), guanine (G), and cytosine (C) just like DNA, but instead of thymine (T) as the fourth letter, RNA uses a base called uracil (U). The other big difference is in the sugar that makes up the nucleotide. Ribose sugar is much less stable than deoxyribose.

So the two genetic molecules don't look much alike. DNA is a long, two-stranded, ladderlike molecule that is very stable and can be handed down unchanged through many generations of cells. RNA is usually just a single strand and can twist itself into complicated 3-D shapes. This versatility is the reason RNA can do lots of jobs that DNA can't do. But RNA also is unstable. Cells constantly break RNA down and replace it. This means they can change their patterns of protein synthesis quickly in response to what's going on around them.

Genes Do the Two-Step

Genes make their proteins (a process called gene expression) in two major steps. The first is transcription, where the information coded in DNA is copied (transcribed) into a molecule of RNA. The ladder of DNA splits in two, and a half-ladder of DNA serves as a kind of mold or pattern for making an RNA copy. The second step is translation: the information now encoded in RNA is deciphered, or translated, into proteins by structures called *ribosomes*.

Tiny Tips

If a word ends in *–ase*, it's an enzyme. But watch out, not all enzymes end in *–ase*.

These steps are carried out with the help of several enzymes, which, you'll remember, are special proteins that speed up chemical reactions.

Transcribing Transcription

Let's explore transcription in a bit more detail. We begin with a piece of double-stranded DNA containing the gene to be expressed. You can visualize this DNA as a kind of zipper. Work starts at the promoter, a regulatory DNA sequence that usually comes before the gene itself. Scientists call this *upstream*.

Nucleotides, also called bases, join together to form the famous DNA double helix. The sequence of one strand is a template or mold for building the other strand because the bases join together in only one way: A with T and G with C.

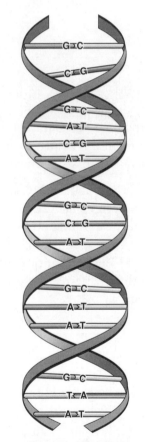

The enzyme RNA polymerase (RNAP) hones in on the promoter and binds to it, which causes the zipperlike DNA to begin to unzip. RNAP moves along the opening zipper, and using the DNA as a template for stringing complementary nucleotides together in an elongating chain, makes complementary RNA copies of the DNA as it goes. The copies are known as messenger RNAs (mRNA). The talented RNAP also proofreads the mRNA it makes, removing mistakes.

When completed, the mRNA is released from the DNA-RNAP structure and now is ready to be translated into protein. The release occurs upstream while transcription is taking place downstream. This is especially true in prokaryotes, where transcription and translation take place simultaneously on the same piece of mRNA. The process of decoding mRNA to produce a particular amino acid chain specified by the genetic code takes place in three steps similar to those in transcription: initiation, elongation, and termination.

Translating Translation

Translation takes place in the ribosomes, which are complex cell structures made of proteins plus a particular kind of RNA called ribosomal RNA (rRNA). Ribosomes are sometimes likened to factories and sometimes to workbenches. Cells possess many thousands of them; for example *E. coli* is said to have more than 15,000.

Now's the time to explain how the genetic code works. As we noted earlier, DNA and RNA are simply sequences of nucleotides that spell out the genetic code. A group of 3 nucleotides (known as a codon or sometimes a triplet) specifies which one of 20 amino acids should come next in a polypeptide that will form part of a particular protein.

For example, the RNA code for the amino acid glutamic acid is GAG (guanine, adenine, guanine) and the code for histidine is CAC (cytosine, adenine, cytosine). But the genetic code is redundant, so almost all the amino acids can be specified by more than one code. Thus GAA (guanine, adenine, adenine) also codes for glutamic acid and CAU (cytosine, adenine, uracil) for histidine. Recall that RNA uses U instead of T for its fourth "letter."

Thus a structural gene is a string of codons that directs the building of a specific protein by specifying the exact order of the amino acids that are its components. The gene also contains codons that start and stop the construction.

Translation requires a different kind of RNA, a small molecule called transfer RNA (tRNA). tRNAs are complementary to the mRNA codons. Each tRNA links to its matching amino acid and carries (transfers) it to a ribosome. There the amino acids are put together in the order specified by a gene's mRNA to make a particular protein.

Making a eukaryote protein is a two-part process, transcription followed by translation.

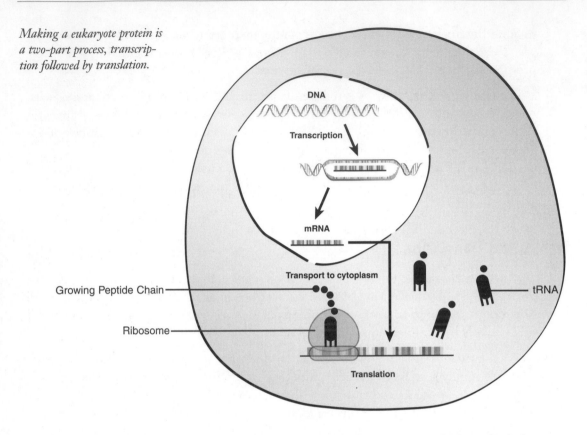

Making a eukaryote protein is a two-part process. During transcription, the information in a gene's DNA is passed on to mRNA in the nucleus, which takes it out into the cell's cytoplasm. Then comes translation: the ribosome deciphers the information about the amino acid sequence encoded in the mRNA. tRNA then builds a protein by stringing together the amino acids in the proper order. Here's how. The ribosome links to a gene's mRNA and runs along the mRNA sequence, matching each mRNA codon to its complementary tRNA codon and adding the tRNA's amino acid to the lengthening peptide chain. When the ribosome encounters a stop codon (a codon that does not have a comparably charged tRNA carrying an amino acid), it stops, letting go of the protein and its mRNA. At this point, the protein folds into its proper configuration (usually) and begins carrying out its function in the cell.

Differences Between Prokaryotes and Eukaryotes

Prokaryotes do both their transcribing and translating in the cytoplasm. Translation starts while transcription is still going on.

As you might expect, in eukaryotes this process is more complex in several ways. For one thing, they work with three major RNA polymerases, not just one.

Eukaryotes, remember, keep their DNA in a nucleus, so that's where transcription takes place, too. That offers opportunities for regulation of gene expression not available to prokaryotes. mRNA isn't released from the nucleus until the cell is ready to begin translating it into protein, which is one way of controlling the time of protein manufacture.

Another example: Before translation, eukaryotic mRNA is edited by splicing out noncoding sequences and putting coding sequences together in different configurations. We'll explain noncoding sequences in a moment, but we want you to know that these splicing variants can make somewhat different proteins that have different functions. This flexibility makes it possible for eukaryotes (especially big, complicated eukaryotes like people) to use a single gene for a number of different jobs.

During these steps, remember, the cell makes several types of RNA:

◆ Messenger RNA (mRNA) gets translated into protein.

◆ Ribosomal RNA (rRNA) helps build ribosomes for making the protein.

◆ Transfer RNA (tRNA) carries amino acids to the protein under construction.

DNA transcription of the genes for mRNA, rRNA, and tRNA produces precursor molecules that are further modified before they are ready to begin working. Some of these modifications are under the control of a fourth type of RNA, small nuclear RNA (snRNA).

Don't Call It Junk

Genomes appear to be full of miscellaneous mysterious sequences, some of which might be garbage. Researchers were astonished when they realized that most organisms contained long stretches of DNA that didn't seem to do much of anything.

At first they called it junk DNA. There are several hypotheses about the origin and function of junk DNA. It might, for example, represent bits of genes an organism has discarded in its evolutionary past. Or perhaps it's shards of unsuccessful invading viruses. Scientists also hypothesize that having a lot of useless DNA around to mutate ensures that most mutations will not be harmful.

As they plunge further into genomes, however, researchers have learned that at least some of the junk does things like perform regulatory functions, for example. "Junk" turns out to be the wrong word for this genetic material. So it's now officially called noncoding DNA because, although it doesn't make proteins, some unknown proportion of the former junk is not junk at all.

Eukaryote genomes tend to be heavy with noncoding DNA, sometimes astonishingly so. Consider yours, for example. The human genome is more than 98 percent junk—uh, sorry—noncoding DNA. Microbial eukaryotes usually have small genomes, but even they contain a lot of noncoding DNA. Yeast has a genome that is 30 percent noncoding. Prokaryotes, with their even smaller genomes, usually have less. The bacterium *E. coli*'s genome is about 11 percent noncoding.

Transposons

One sort of DNA that has sometimes been called junk is transposons, a short-hand term for transposable elements. A transposon is a chunk of DNA that can move around in a genome, often simply jumping to another part of the same chromosome or to a plasmid. For that reason transposons are also called mobile genetic elements and even jumping genes.

The reason transposons have sometimes been classified as junk is that in their natural state, scientists don't find many good things to say about them. However, scientists have turned them into tools for research, and this is the reason we now have multicolored Indian corn. We'll talk about some of that research later in this chapter.

When transposons move around, they can cause mutations by inserting themselves into the middle of a gene. That can disrupt the gene's function and cause disease. Transposons often make copies of themselves as well, so they can alter the amount of DNA in a genome, sometimes dramatically. Eukaryotes are full of transposons. Nearly half of your DNA is estimated to be transposons or their remains.

Transposons usually contain at least one gene. That's the gene coding for transposase, an enzyme they need to open up places for themselves at target sites elsewhere in the cell's genome. A transposon also sometimes contains a gene for copying itself, but more often it simply hijacks the cell's copying mechanisms.

Genes for antibiotic resistance also can be carried around and copied by a transposon, which is how they get into plasmids and get passed along to other organisms. Sometimes transposons contain promoters or stop codons, which means they can switch nearby genes on and off with unpredictable results.

There are two main kinds of transposons: Class I and Class II. Retrotransposons, Class I, make use of transcription, copying themselves to RNA. But instead of moving on to conventional translation, a retrotransposon uses an enzyme called reverse transcriptase (often coded by the transposon itself) to turn the RNA copy back into DNA. The new DNA then gets pasted back into the genome in a different spot. Class II, autonomous transposons, code for a transposase that helps them cut themselves out of one spot in a genome and insert themselves into another.

Gene Sequencing: The Alphabet of Life

Genetic sequencing means discovering the order of bases (nucleotides) in a gene or any other piece of DNA, including the entire genome of an organism. When scientists know the exact sequence of a gene, they can often figure out what protein it makes. They can also discover mutations that keep it from working correctly.

A genome sequence is a huge leap toward understanding how an organism is built. But the sequence is just the first step; a lot of work remains. After they have sequenced an organism's genome, scientists still must figure out the following:

- What each gene does
- What each of the proteins made by the gene are doing
- How the genes and proteins all work together to create life

You may have heard of the Human Genome Project, the massive international program to learn our own complete DNA sequence. Researchers published the close-to-final human sequence in 2004. They are now busy trying to figure out what all of our 25,000 genes do, and how they relate to each other. That's a job that will last for most of the twenty-first century.

But there are also tons of genome projects for other organisms, large and small. Hundreds of microbe genome projects have been completed and hundreds more are in the works.

Why Sequence Microbe Genomes?

Well, obviously, understanding the genomes of pathogenic microbes will help keep us healthy, and keep the plants and animals we depend on healthy, too. Those genomes will help scientists find pathogens' vulnerabilities and hunt down new prevention and treatment strategies. Already many pathogen genome projects are complete. We now have DNA sequences for microbes that cause pneumonia, tuberculosis, gonorrhea, meningitis, malaria, and salmonellosis, to name just a few.

Sequencing microbial genomes also will help keep the planet healthy. There are genome projects on microbes involved in climate change, microbes that may generate new sources of energy from methane or hydrogen, and microbes that can clean up toxic wastes and improve industrial processes. Microbial genome information also can help assure safer food and water.

As the number of available microbial genome sequences grows, microbiologists are doing more and more work in a newer field called comparative genomics. Comparing microbe genomes helps them …

♦ Work out who is related to whom and devise far more accurate classification systems.

♦ Identify previously unknown microbes.

♦ Figure out the lifestyles of the many microbes scientists haven't yet been able to grow in their labs and study up close and personal.

Here's just one example of the utility of comparing genomes. Three different parasitic protozoa transmitted by three different insects cause three different tropical diseases that affect millions. None can be prevented or treated effectively. *Trypanosoma cruzi* causes Chagas' disease in Latin America, *Trypanosoma brucei* causes African sleeping sickness, and *Leishmania major* causes the skin disease leishmaniasis around the world. Despite their surface differences, genome comparisons show the three protozoa share a core 6,200 genes. These common genes are expected to offer targets for new ways of combating all three diseases.

Recently scientists have begun to tackle hugely ambitious metagenome projects. They are sequencing the DNA of entire mini-ecosystems, such as all the microbes that inhabit a bit of soil or a bit of the human body. Examining these microbial communities in DNA detail is turning up many never-before-seen creatures, and is expected to reveal lots more. It also will generate practical outcomes like new medicines and new ways of beating pollution, plus other benefits not yet dreamed of.

Dazzling Diversity

One fascinating result of all this work has been to teach us that microbes are even more different from each other than microbiologists had previously imagined. You may think you are not at all like other mammals, yet 96 percent of a chimp's genes are pretty much the same as yours, and so are 85 percent of a mouse's. Yet in the microbe world, comparatively few genes, except for some involved in protein-making and basic cell housekeeping, are widely shared. Even heterotrophs don't seem to have similar basic metabolism genes handling their processing of organic food.

Microbial genome projects have also revealed that new microbe species often arise by getting their new genes not from their parents, but from other microbes. This horizontal (or lateral) gene transfer is quite different from the typical origin of a new species in larger organisms, which occurs via vertical transfer of new mutations from mom and dad. Researchers say that some microbes appear to have acquired as many as one in four of their genes horizontally from other types of creatures.

These revelations demonstrate one reason scientists find it so tough to figure out how microbes are related. Microbe genome projects also have cast doubt on standard classification systems, especially one based on the presence or absence of just a single gene, as with a commonly used system for classifying prokaryotes.

Traditional genetic classification systems assume that genes are handed down from one generation to the next and that you can track the ancestry of an organism by examining its genes. But what if it slurped up lots of its genes from nearby microbes of completely different types? For those new genes, the organism's ancestry is irrelevant.

> **Little Did You Know**
>
> Microbes don't exchange genes only with each other. They engage in lateral transfer to bigger organisms, too; people, for instance. Researchers recently showed that the human genome contains more than a hundred genes that originated in bacteria.

Consider one example from two different strains of the well-studied bacterium *E. coli*. The public has gotten to know *E. coli* 0157:H7 well because this bug has caused devastating food poisoning from fast-food burgers and other foods. Its genome turns out to be 20 percent bigger—that's about 1,000 genes more—than the genome of *E. coli* K12, a benign strain and favorite model organism. It also appears to contain genes from other bacteria—for example, a toxin gene that is similar to the toxin gene in *Shigella dysenteriae*.

Genetic Engineering of Microbes

Genetic engineering (genetic manipulation or modification) means deliberate tinkering with an organism's genes. It can involve inserting new genes, even from an entirely different kind of organism or virus. Or it can mean changing or removing existing genes. Scientists usually call genetically engineered organisms *transgenic*.

The goal is to alter or even shut down the proteins the modified life-form produces. Why? Scientists working in a field called functional genomics may want to discover what a particular protein does, and eliminating it is one way to find out. Others are hoping to make organisms more efficient, more useful, or healthier.

The first commercial use of modern genetic engineering techniques represented a sort of birth day for today's biotechnology industry, which we talk about in Chapter 19. That happened in 1982, with the transfer of a gene for making human insulin into bacterial plasmids. The engineered plasmids entered other *E. coli*, which turned them into pharmaceutical factories making large amounts of human insulin for treating diabetes.

Genetic Engineering with Microbes

As we learned in Chapter 2, bacterial plasmids are one kind of vector for genetic engineering of microbes. Viruses are another kind of vector for genetic engineering, which makes good sense since viruses already do a good job of infecting cells.

Viral vectors are a major method for genetic engineering of plants and animals. To bestow a new gene on an organism, the researcher attaches the gene (or genes) of interest to a virus that can carry foreign genes, and then induces the engineered virus to infect some of the organism's cells. That's the plan anyway, but it has often proved frustratingly hard to do.

Gene Therapy for People

Gene therapy is the form of genetic engineering aimed at treating disease. The idea is to transfer genes into the cells of a patient so that the new genes will lessen the disease and perhaps even cure it. Typically the point of gene therapy is to produce working versions of defective proteins.

In the past, the targets of gene therapy have been genetic diseases, where the aim is to supply healthy versions of malfunctioning genes. Recently researchers have become more interested in chronic diseases like cancer and rheumatoid arthritis.

Gene therapy has been promoted for decades as an excellent way to deal with disease. A few experimental therapies have been believed to be promising before serious problems brought them to a halt. Some therapies have proved dangerous or even fatal to patients. Researchers are hopeful, though, that new methods will overcome existing technical problems and move gene therapy into mainstream medicine eventually.

When scientists are researching human gene therapy, the "good" human gene is inserted into a mouse retrovirus that has been deactivated. The virus attaches to a patient's cell in a laboratory dish and discharges its genetic material into the cell. If all goes well, the therapeutic gene integrates into the patient's DNA, replacing the defective gene, and the genetically engineered cell is then transferred back to the patient, where it performs like the normal gene.

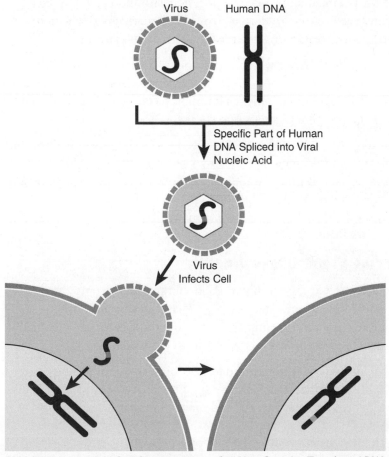

A schematic illustration of how human gene therapy is done.

Researchers experiment with a number of microbial vectors, such as naked plasmid DNA. But three viruses—retroviruses, adenoviruses, and adeno-associated viruses—have been the chief gene therapy vectors to date. All have disadvantages.

Retroviruses are RNA viruses, some of them pathogenic. Even when they are not directly pathogenic, however, retroviruses can cause disease because they integrate unpredictably into the patient's genome. Gene therapists at first had enormous success at treating severe immune deficiency (SCID), better known as the Bubble Boy disease. SCID kids at last were able to live near-normal lives in the real, germ-filled world. But three of the patients have developed cancer because the vector's insertion points disrupted their DNA.

A subclass of complex retroviruses known as lentiviruses is getting attention because they are long acting and can infect, and thus transport genes to, non-dividing cells like nerve cells. But lentivirus vectors are derived from the lentivirus HIV, which causes AIDS, so scientists are investigating their safety with great caution.

Adenoviruses are DNA viruses, many of which cause not-very-serious respiratory infections, otherwise known as the common cold. But even harmless versions of these viruses used as vectors can stimulate the patient's immune system to attack and keep them from delivering their therapeutic gene cargo.

An adeno-associated virus (AAV) requires an adenovirus nearby for reproduction. But on its own, an AAV inserts its DNA consistently into a certain predictable place in the human genome. And because it's not pathogenic, an AAV does not attract hostile interest from the immune system. An AAV's problem is that it can only transport a small amount of foreign DNA, which rules it out for many diseases.

Designer Genes and Making Organisms to Order

Some of the same principles researchers apply to gene therapy can be employed to create genetically engineered plants and animals. There is one crucial difference, however: in these organisms, scientists engineer the reproductive cells (often called the germ line) so that gene alterations are passed down to descendants.

Already they have built hundreds of different model organisms for study. The genes of laboratory mice have been engineered to display symptoms of scores of human diseases, the goal being to figure out how to treat, and maybe prevent, the ailments.

Plant genetic engineering has aided scientists to study plants and also created plant "factories" that churn out molecules of commercial interest, especially to the pharmaceutical industry. And plant genetic engineering has become crucial to agriculture. The vectors plant genetic engineers most rely on are Ti plasmids, derived from the soil bug *Agrobacterium tumefaciens*, which causes plant tumors.

Tiny Tips

Genetically engineered mice are often called knockout (KO) mice when a gene has been knocked out (sometimes by inserting a transposon to disrupt its function). When a gene has been added, they are sometimes known as knock-in mice.

Genetic Engineering Controversies

People have been breeding animals and plants deliberately for thousands of years, but today's genetic engineering techniques raise questions that we'll continue to wrestle with. One biggie is safety.

- Suppose the pest-resistant plant transfers its genes to other plants, especially weeds?

- Is genetically engineered food wholesome and healthful?

- Can gene therapy be made both effective and safe even though some gene therapy patients have died?

Another area of concern, particularly important in agriculture, is economics. If farmers use seed that has been genetically engineered so that the resulting plants cannot reproduce, that means they must buy new seed every year instead of seed they have gathered from the previous year's crop for free. How big a financial burden is that, especially in very poor countries?

People also are worried about applying genetic engineering to ourselves. Is it right to want to invent a new and improved human? Is it safe? Will it give some people unfair genetic advantages?

Improvements in genetic engineering technology eventually may help with some of these questions. But for many of them, there are no right or wrong answers, only opinions that differ strongly. In other words, they'll continue to be discussed and debated for a long time to come.

The Least You Need to Know

♦ Earth's billions of life-forms have genes that are made of the chemical DNA, which is just coded information, a kind of instruction manual for a cell.

♦ Another kind of genetic material is RNA. RNA, probably the ancestor of DNA, actually carries out DNA's instructions for making proteins.

♦ There are hundreds of microbe genome projects aimed at finding out the exact order of the chemicals that make up DNA.

♦ The genes of microbes can be engineered to make new kinds of microbes, and microbes can also be used as vehicles for transferring new DNA to other organisms.

♦ This genetic engineering means that organisms can be changed to make useful products, and also provides hope for curing diseases.

Chapter 7

Cell Structure and Cell Structures

In This Chapter

- ◆ Microbial cells and their contents
- ◆ Cell structure—and cell structures
- ◆ Prokaryote cell membranes and cell walls
- ◆ Peptidoglycan, an important microbial molecule
- ◆ Eukaryote cell membranes and cell walls

All living things are made of cells. Complex creatures such as fruit flies and people are constructed of billions, sometimes trillions of cells, each one specialized to perform a particular set of tasks.

Except for viruses and prions, which are not alive and therefore possess no cells, most microbes are just a single cell. But that one cell does a lot. In this chapter we'll tell you a little about what microbial cells do and why it matters.

Cells and Cell Structure

Cells do the basic work that keeps an organism going. Cells absorb food and transform it into energy. They provide structure to the organism and also contain the genetic instruction manual for performing specific tasks.

What Microbial Cells Have in Common

Cells are tiny bags filled with a fluid called cytoplasm. Cytoplasm is mostly water, but it also contains salts and thousands of proteins and protein fragments. The proteins busily carry out chemical reactions that help cells generate energy, grow, and make new cells.

Cytoplasm also contains various cell structures. In prokaryotes, these include …

♦ Inclusion bodies, used for storage of energy, carbon compounds, and inorganic substances.

♦ Ribosomes, the cell's protein factories.

♦ The nucleoid, home to a prokaryote's DNA.

As you might expect of a storehouse, inclusion bodies are often surrounded by membranes that secure their contents—but not always; some float around in the cytoplasm. Inclusion bodies carry out a variety of tasks depending on the organism. For example, cyanobacteria and other photosynthetic bacteria often possess gas vacuoles. These balloonlike inclusion bodies keep the bugs afloat so they can capture sunlight to make energy.

Not all inclusion bodies are for storage. Some provide a magnetic guidance system that help bacteria (and birds and aquatic organisms) navigate via Earth's magnetic field. Called magnetosomes, these structures contain iron particles that serve as magnets, drawing organisms downward and toward Earth's poles.

The protein-producing ribosomes don't look like much, but most cells are stuffed full of thousands of them, and they are quite complicated. Ribosomes suspended in cytoplasm make proteins that will stay in the cell. Proteins destined to be sent to the outside world are manufactured in ribosomes loosely attached to the cell membrane. (We discuss cell membranes in a moment.)

Ribosomes of prokaryotes are a bit smaller (about 15 nm × 20 nm) than those of eukaryotes (about 22 nm diameter). But both are made of one large and one small subunit. We explore the structure of ribosomes and what they do more in Chapter 8.

In the meantime, note that in addition to their central role in protein production, bacterial ribosomes are important for another reason. Some classes of antibiotics attack them, crippling a bacterial pathogen's protein production. The aminoglycoside antibiotics (streptomycin, neomycin) and the tetracyclines bind to the small subunit, while a few others latch on to the large subunit.

You might say that the nucleoid defines the essential difference between prokaryotes and eukaryotes. Eukaryotes carry most of their genetic material around in the nucleus, a structure that prokaryotes do not possess by definition. Instead, prokaryote genetic material resides in an ill-defined cell region with variable boundaries that also has varied names. We call it the nucleoid here, but you will also encounter the terms nuclear body, nuclear region, and chromatin body.

Two other structures are common (but not universal) in microbes. Both have been mentioned before and will be again. One is the flagellum (plural: flagella). That's an organelle, often said to be whiplike, that microbes use to scoot around. This structure gives them all mobility, but is quite different in archaea, bacteria, and eukaryotes.

> **Little Did You Know**
>
> Eukaryote cells also usually contain specialized structures called organelles, little organs that are enclosed by membranes. The organelle story is entrancing, but it's really part of the story of symbiosis, so we save that for Chapter 9. But here's a teaser: organelles, it turns out, are microbes, too! At least many started out that way.

Bacteria employ a spiral-shaped protein that is actually a hollow tube. It possesses a protein engine situated at the junction with the cell membrane and is driven by ions (usually hydrogen ions) that stream across the membrane. For an explanation of ions, see Chapter 5. Some bacteria species have just one flagellum; others have many.

Flagella in archaea look a bit like bacterial flagella but use a different engine mechanism that apparently does not involve ions. Also, their flagella rotate together as a single mechanism, whereas bacterial flagella rotate independently of each other. There are also structural differences between the two.

Eukaryote flagella are usually called cilia (singular: cilium). That's appropriate since they aren't at all like prokaryote flagella in structure, even though their main function, mobility, is the same. A cilium is a collection of nine fused pairs of tiny hollow tubes (microtubules) enclosing two single microtubules. Researchers think cilia may have other functions, too, such as communication between cells.

Spores are reproductive cells and will be discussed more elsewhere, especially in Chapter 8. There are many kinds of microbial spores, both prokaryote and eukaryote.

Spores are often compared to seeds, except they contain no stored food. They are ubiquitous and tend to be very sturdy structures, able to withstand dehydration and other environmental stresses, surviving and persisting for a long time—sometimes decades and even longer—until conditions are right for germination.

Membranes and Walls

Each cell is surrounded by a thin membrane, or sometimes two, depending on the organism. Membranes are essential for life. Cells must hang on to the stuff they need to survive and keep other stuff outside. So they need to put a barrier of some kind between themselves and the rest of the world, a fence with gates for managing traffic in and out. Cell membranes are studded with protein tunnels that work like channels and pumps, opening and closing the gates to let molecules in and out of the cell.

This membrane fluidity lets the cell carry out its basic physiology, taking in food and dumping its garbage. The changeable membrane also helps cells communicate with the world outside, including other cells.

Many kinds of microbes also possess a protective cell wall that lies outside the membrane and is somewhat more rigid. But some eukaryotes, especially some protozoa, have extra stiff cell membranes and so can get along without cell walls.

Eukaryotes make far more use of membranes than prokaryotes do. Additional membranes help keep the specialized functions of these more complex creatures separated and organized. Membranes surround their genetic material, creating the nucleus, which, you'll recall, is the cell feature that distinguishes eukaryotes from prokaryotes. Whether eu- or pro-, cells always contain the genetic material DNA. There's a lot to say about microbial genomes, so we've saved that for Chapter 8.

Eukaryote cells also contain scaffolding, the cytoskeleton. There are some hints that perhaps prokaryotes have cytoskeletons, too, but that's not certain yet. The cytoskeleton acts a bit like muscle to move parts of the cell around. This essential structure also helps cells preserve their shapes and keeps organelles organized and securely attached.

How Many Cells Make a Microbe?

Many microbes—for example, bacteria—are unicellular, meaning they consist of just one cell. This bacterial cell is usually much smaller than any cell in your body. Often species of bacteria live together in colonies, but while each cell in this mass appears identical to every other, it can live independently, too.

Some microbes are multicellular, like the mold on bread and other fungi. In Chapter 3, we discussed how mold cells (for example, cells of the bread mold *Neurospora crassa*) join end-to-end to form threads of cells known as hyphae and clot together in a tangled mass called mycelium.

Bacterial Cell Walls and Membranes

As we said previously, cell membranes regulate which molecules can enter and leave a cell. Membranes are essential for life, and all organisms possess them. In addition, nearly all prokaryotes possess cell walls. The small group that is the exception to this rule includes mycoplasmas and a few archaeans.

Bacterial Cell Walls and Osmosis

Most prokaryote cells are enclosed by a strong wall. This wall gives them their shape and also protects them from toxins and other potential sources of damage. One of the wall's main functions is to prevent cells from undergoing osmotic lysis—that is, bursting open as a result of pressure from osmosis.

Osmosis is the diffusion of liquid from a weak solution to a strong solution across a semipermeable membrane. A semipermeable membrane is only porous enough to permit a solvent like water to pass through, leaving dissolved solids like salt behind.

The aim of osmosis is to equalize the solution strength on either side of the membrane. The watery environment outside bacteria usually contains more dissolved solids than the cytoplasm within. So if a bug was protected only by a membrane rather than having a strong cell wall, water would tend to flow into it because of osmosis.

Eventually, water pressure would cause the bacterium to pop like an overfilled balloon. Cell walls keep that potentially dangerous fluid outside.

In bacteria, the wall is strong and rigid and resists osmosis mostly because it contains peptidoglycan, pronounced pep-ti-doe-GLY-kan and abbreviated PGN. PGN has been called the most important molecule bacteria possess, so let's take a closer look.

Why Peptidoglycan Matters

PGN (murein) is a huge molecule made of many interlocking chains of identical subunits that combine carbs with peptides (which are, you'll recall, snippets of proteins). The subunits possess a long backbone of two alternating sugars (polysaccharides) festooned with short chains of amino acids.

These amino acids vary a bit, depending on bacterial species. Three of the amino acids are not in any protein. They protect the molecule against enzymes that break up proteins.

The long rows of sugar chains in peptidoglycan are coupled to each other by identical peptide cross-links made of four amino acids. This cross-linking turns the polysaccharide chains into an immensely strong lattice.

Peptidoglycan is a big molecule made of repeating sugars that form long parallel chains, linked together with peptides into a strong structure similar to a chain-link fence.

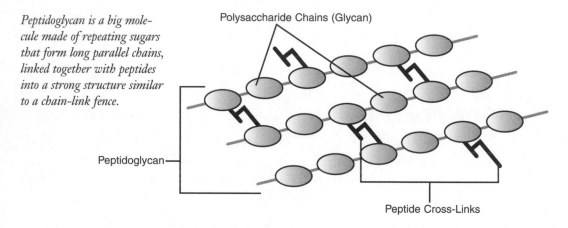

Polysaccharide Chains (Glycan)

Peptidoglycan

Peptide Cross-Links

Microbiologists have compared this lattice to a chain-link fence surrounding the bug. They think of it as one immense molecule that functions like a shell or bag to contain the organism. The fence is many layers thick, but the thickness depends on the organism.

The cross-links can vary between bacterial species. Staphylococcus PGN has many cross-links, but *E. coli*, with fewer, possesses a more open PGN meshwork.

There are other variations in PGN among bacterial species. But in all of them, the cell wall is hugely important. That is especially true for bacterial pathogens. Many antibiotics work by attacking the enzymes that help construct PGN in the cell walls of disease bacteria.

For example, penicillin prevents the enzyme that forms cross-links between amino-sugar chains from doing its job. The result: a weak cell wall, osmosis, and a busted bacterium. Other groups of antibiotics work this way, too, including the cephalosporins, the carbapenems, the monobactams, and the glycopeptides.

Tiny Tips _____

Earlier we said that the mycoplasmas are the only group of bacteria that don't possess cell walls. So how do they fight off osmotic lysis? By actively pumping sodium ions out of the cell. Mycoplasmas cause some kinds of pneumonia and genitourinary infections. Penicillin doesn't work against them because they don't have PGN that can be attacked. Detergents, however, will cause mycoplasma cells to rupture.

In Chapter 2, we said that bacteria could be divided roughly into two groups on the basis of Gram staining. After staining, gram-positive bugs appear purple and gram-negative ones look pink. These colors are due to differences in the two groups' cell walls, differences chiefly in the amount of peptidoglycan they contain.

The cell walls of gram-positive bugs contain a thick deposit of peptidoglycan. This deposit can be up to 80 nm in size and composed of dozens of layers, which makes the walls particularly strong.

The cell walls of gram-negative bacteria contain only a small amount of peptidoglycan. It is always less than 10 nm in size and sometimes as thin as one layer.

Bacterial Cell Membranes

This narrow peptidoglycan layer of gram-negative bacteria is overlaid by a membrane known as the outer membrane or lipopolysaccharide layer (LPS). So this thin cell wall is sandwiched between the outer membrane and the regular cell membrane all organisms possess.

In organisms that also possess cell walls, a category that includes almost all life-forms, the cell membrane is often called the plasma membrane. That terminology distinguishes it from the outer membrane, which is the layer just outside the wall.

With an electron microscope it is possible to see a space between the plasma membrane and the cell wall. You can also see the space between the plasma membrane and the outer membrane in cells that have outer membranes. The space contains peptidoglycan and other molecules such as proteins.

In prokaryotes, cell membranes must do work that in eukaryotes is performed by organelles inside the cell, such as photosynthesis and synthesis of essential molecules. From the microbiologist's point of view, the most important difference between types of bacteria may lie in their cell walls and membranes. Oftentimes membrane chemistry is so characteristic of a specific species that it can be used as a way to identify just which bug is causing a problem.

Gram–positive and gram–negative bacteria contain different amounts of peptidoglycan in their cell walls, as shown in these schematic drawings. Gram–negative bacterial cells also possess an additional outer membrane.

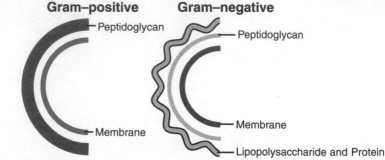

As we discussed in Chapter 5, cell membranes are mostly proteins and lipids. The plasma membrane (which is what the prokaryote cell membrane is often called) is composed of a double layer of lipids studded with proteins.

Some of the membrane proteins form tunnels or portals through the membrane to let in essential molecules and get rid of waste. Some of the proteins sit at the surface. These surface proteins have various functions, one of which is to act as receptors for signaling molecules. Sometimes surface proteins are decorated with carb bits whose function is not yet known.

Small proteins are lodged inside the membrane in among the lipids. Their function also is not yet known. The lipid level often is called the lipid or phospholipid bilayer (pronounced BY-layer).

Recall that lipids are just fats, and they are insoluble in water. The molecules of fat in the lipid bilayer have one end with no electrical charge so it is often called the nonpolar end (or sometimes the nonpolar tail end). This nonpolar tail is hydrophobic (water-repelling). It is lodged deep inside the membrane.

The other end does have a charge so it is called the polar end (or sometimes the polar head end). The polar head is attracted to water (hydrophilic). It sits at the surface on both sides of the membrane.

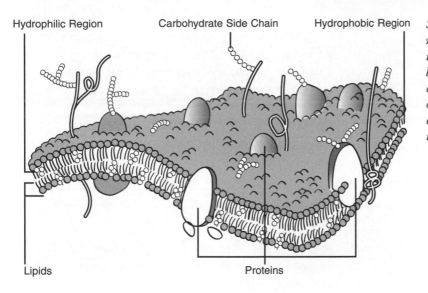

Hydrophilic Region   Carbohydrate Side Chain   Hydrophobic Region

Lipids   Proteins

Schematic drawing of a microbial plasma membrane, illustrating its phospholipid bilayer, with the hydrophobic ends of lipids facing each other and the hydrophilic ends facing the outside and inside of the cell.

Thus the lipid bilayer consists of lipids stacked in groups of two. Their nonpolar tails face each other within the membrane and tend to associate with each other.

This double row of lipid molecules forms a protective barrier guarding the cell contents. Their polar heads face outward in opposite directions, one facing toward the watery cytoplasm within the cell and the other at the outer surface of the membrane, facing the watery world outside the cell.

Despite this complex structure, outer cell membranes aren't very thick, less than 10 nm, which makes them visible only when viewed under an electron microscope. The outer cell membranes also have long-chain sugars on the outside of the membrane, some of which are unique to bacteria.

Prokaryotes also possess internal membranes that segregate some interior functions from each other. These membranes can form when the plasma membrane folds into the cell, making a kind of pocket that is called invagination.

Archaeal Cell Walls and Membranes

Archaea can be gram-positive or gram-negative, but their cell walls don't contain peptidoglycan. Instead an archaean's cell wall, while often fairly rigid and protective, is full of proteins, glycoproteins, and polysaccharides.

Wee Warnings

Some archaean cell walls contain a complex molecule called pseudomurein that has been called a PGN, although it has quite a different structure. Penicillin does not work against archaean cell walls.

Membranes of many archaeal cells are simpler than bacterial plasma membranes; they contain only a single layer of lipid molecules. Archaeal membrane lipids differ from lipids in bacteria and eukaryotes and are connected in diverse ways. Many are nonpolar, which means they can be combined in several ways to make membranes that vary in rigidity and thickness.

Eukaryote Cells

Eukaryotes may be more complex than prokaryotes in many ways, but their cell walls are simpler, and they contain no peptidoglycan. Their function is the same as prokaryote walls, however, to maintain shape and prevent osmotic lysis.

Some kinds of eukaryotes lack cell walls altogether. Membranes play a larger role in all eukaryotes, often separating functional parts of the cell from each other. One prominent example is the nucleus, surrounded by a cell membrane that keeps DNA inside.

Eukaryote Cell Walls

If eukaryote cell walls are not made of peptidoglycan, what are they made of? The answer is polysaccharides, complex, insoluble long chains of simple sugars. Just which polysaccharide depends on the type of eukaryote.

- In algae, many fungi, and plants, the chief component of cell walls is cellulose, which is essentially wood fiber.

- Molds often have cellulose in their cell walls, too, but the main ingredient is chitin, a tough material also found in the exoskeletons (shells) of insects.

- Protozoans, some fungi, and animal cells have no cell walls.

Let's take these in turn. Cellulose accounts for about half the weight of a plant cell and is said to be the most abundant polymer on Earth. It typically is composed of thousands of glucose molecules strung together to form long chains. Dozens of these chains gather together in rows known as microfibrils. These are set in a mixture of proteins and other polysaccharides and cross-linked to form a strong, rigid matrix.

In algae, the cellulose is mixed in with polysaccharides. Groups of algae often have characteristic polysaccharides, which makes it easy to identify them. The crystalline shells of diatoms also contain silica, which is plain old sand.

Cellulose is almost impossible to break down except with the help of certain bacteria that can digest it. Bacteria with this talent reside in the guts of ruminants such as cows and in the termite gut—but usually not in yours.

Chitin has a chemical structure similar to cellulose, with one difference that permits additional hydrogen bonding between each link in the chain. Thus chitin is stronger and more rigid than cellulose, as you can tell when you compare wood fiber with insect shells.

Eukaryotes Without Cell Walls

Calcium shells are a cell-wall substitute in some protozoa, providing protection and controlling osmotic pressure. Others possess a flexible, or sometimes rigid, covering known as the pellicle, which also performs those functions. Pellicle composition varies depending on the species. The euglena pellicle is made of strips of protein that spiral around the cell. Some other protozoa are surrounded by many tiny balloonlike air sacs called alveoli.

But many freshwater protozoa handle the water problem in a different way. They maintain water balance via a special compartment near the membrane. Called a contractile vacuole, this compartment fills with fluid and then contracts to squeeze excess water out of the cell.

Animal cells possess plasma membranes that protect them somewhat and handle water balance, but they lack cell walls.

Without cell walls, they also had to develop additional ways of protecting themselves. One strategy was formation of an extracellular matrix made largely from polysaccharides and a sturdy protein called collagen. The matrix lies outside of cells but binds them together.

The Least You Need to Know

- ◆ All life-forms, including bacteria, protists, and fungi, are made of cells, which perform the basic work of keeping an organism going.

- ◆ Cells are watery bags that often contain complex organelles, and they are always surrounded by membranes that let in food and other essentials and get rid of waste products and potentially damaging molecules.

- ◆ In addition to membranes, many microbes also possess cell walls that keep them from bursting open because of osmosis.

- ◆ An important microbial molecule is peptidoglycan, a complex lattice-like molecule made of protein and sugar that makes the cell wall rigid and protective.

- ◆ Many antibiotics work by attacking the bacterial ability to make peptidoglycan.

- ◆ Eukaryote cell walls don't contain peptidoglycan but are protective because they contain long chains of polysaccharides.

8

Microbial Lifestyles

In This Chapter

- Nutrients microbes must have
- Autotrophs and heterotrophs
- How microbes eat
- How microbes reproduce without sex
- The bacterial growth curve
- Microbes and sex

Microorganisms eat, grow, and reproduce themselves like other living things, even though they do it differently from many other living things, like you. Viruses don't eat, but they do reproduce and grow in number. In this chapter, we explain how microbes accomplish these commonplace tasks.

Microbes Are What They Eat

Like you and other organisms, microbes must take in certain chemical elements to build molecules such as carbohydrates, lipids, and proteins to make energy, and to survive. They all need a lot of elements such as carbon, oxygen, hydrogen, calcium, and potassium, and tiny amounts of trace elements such as zinc and copper.

Tiny Tips

Microbiologists call fussy microbes that insist on certain nutrients or other particular conditions "fastidious."

A few microbes need special molecules for special circumstances. For example, most microbes don't need much sodium, but microbes that grow in salt water often need lots of it.

Some microbes can get the carbon they need from almost any source that contains it, even very unlikely places like rubber. That's one reason rubber deteriorates with time: something is eating it. Other microbes are very fussy about carbon sources and will take in only certain ones.

Microbial nutrition is enormously important to microbiologists because the right food and other conditions are essential in order to grow and study microbes in the lab.

Eat or Be Eaten

As we noted in previous chapters, lots of microbes make their own energy from chemicals and light in a process called photosynthesis. This includes most algae, some bacteria, and some protozoans. These creatures are called autotrophs and are classified as "producers."

Photosynthesis is the most common form of autotrophy, which is why most autotrophs must live near light. A few autotrophs can manage on inorganic chemicals alone. Microbes that live in deep-sea vents grab what they need from dissolved inorganics that drift by.

Autotrophs produce organic compounds by getting almost all their carbon from carbon dioxide. But many microbes, like fungi and some bacteria, must eat other organisms. They are called heterotrophs and are classified as "consumers" because almost all their carbon comes from eating autotrophs or preformed organic compounds.

How Do Microbes Eat?

Microbes obviously have no mouths or stomachs, so how do they eat? The two main methods of microbial eating are diffusion and active transport.

Small molecules, such as oxygen and water, can diffuse easily across the cell membrane and disperse into the cytoplasm. Sometimes, however, the small molecules need some help from active transport. Active transport, used by larger molecules such as proteins and sugars, get help from cell-surface proteins and transfer the molecules inside. The electrically charged molecules, called ions, get propelled in and out with the help of protein pumps in the cell wall.

Eukaryotic microbes can also take in big molecules and even other cells via endocytosis, another form of active transport. The microbe first surrounds the molecule with its cell wall, forming a little sack. Then it pinches off the sack, which floats in the microbe's cytoplasm.

Lysosomes, whose job it is to degrade molecules brought into the cell, merge with the sack, releasing digestive enzymes that break down whatever is trapped within. Thanks to membranes, lysosomes can perform this trick without releasing those enzymes into the cell, which is definitely a good thing because otherwise the enzymes would digest the cell along with the food.

Endocytosis is a favorite technique of protists. It's useful not only for eating but also for dealing with invaders such as viruses. In the latter case, endocytosis is triggered by the viruses themselves.

How to Grow Microbes

In addition to absorbing nutrients and finding a way to make energy, microbes grow, or not, in response to a number of factors. Among them are ambient temperature, pH, and the presence (or absence) of oxygen.

Temperature

As you know very well when you flee from ghastly summer heat and humidity into an air-conditioned room, temperature is an important factor for all organisms. Unlike you, microbes are mostly one-celled and therefore are about the same temperature inside and out, so they must be able to tolerate ambient temperature, whatever it is. Too low and the microbe's molecules won't work properly. Too high and essential molecules will disintegrate, taking the microbe along with them.

But low and high temperatures are relative terms that depend on the individual species of microbe. Species and their internal chemistry have evolved to cope with whatever temperature Earth manages to dish out. As we have already seen in Chapter 2 with hot springs archaea, some prokaryotes can tolerate much higher temperatures than eukaryotes.

Organisms that depend on photosynthesis may like it sunny, but even so they are not crazy about high temperatures. And just recently researchers discovered a vast mat of bacteria that have been living, apparently happily, under a 600-foot-deep ice shelf in Antarctica for at least 10,000 years.

pH: The Basics of Acid and Alkaline

Each microbe species needs a particular range of *pH* to grow at all, and a specific pH to grow well. pH measures the concentration of hydrogen ions in a fluid, usually water.

def•i•ni•tion

pH stands for potential of hydrogen, and it's the logarithm of the reciprocal of hydrogen-ion concentration in gram atoms per liter. Aren't you glad you asked?

The measure is expressed on a scale from 0 to 14. But the numbers are a bit confusing because a low number means lots of hydrogen ions and a high pH number means fewer hydrogen ions.

An easier way of thinking about pH is that it's a measure of how acid or alkaline a solution is on a scale of 0 to 14. Seven is neutral; anything below seven is acid and anything above seven is alkaline (sometimes called basic).

Bacteria and protozoa tend to like a near-neutral pH, somewhere between 5.5 and 8.0. There are, of course, exceptions, such as *Helicobacter pylori*, which lives in the human stomach where the pH is about 2. Fungi and algae mostly like their environment a bit acid, from 4.0 to 6.0, and there are some archaea that can grow at a pH about as low as it can get, near 0.

As always when we are speaking of microbes, there are exceptions, and some microbes can adjust their pH tolerance a bit when their environments change. A lot of microbes can even control their environments, changing the surrounding pH to suit them better by releasing waste products that are acid or basic.

Oxygen, or None

We learned in Chapter 1 that the microbial world can be divided into aerobes and anaerobes, those that need atmospheric oxygen and those that find oxygen poisonous. Guess what? There are exceptions!

Almost all multicellular creatures need oxygen: they are often called obligate aerobes. Some microbial aerobes cannot tolerate a "normal" level of atmospheric oxygen, that is, 20 percent. These are microaerophiles, and they need low oxygen levels, under 10 percent.

And there are different kinds of anaerobes, too. Obligate anaerobes are, well, obligated to stay away from oxygen or they will die. Facultative anaerobes don't need oxygen, but they like it and do better when it's around. Some other anaerobes don't care either way: oxygen, no oxygen, whatever. Obviously flexibility like this is an adaptive advantage.

The Sex Lives, or Lack Thereof, of Microbes

You may have thought variations in human reproductive and sexual behavior were weird, but microbes are weirder, trust us. We talked just a bit about microbial reproduction in Chapters 2 and 3. It's an incredibly complex subject, and we're only going to be able to hit a few highlights here.

Reproduction Without Sex—Let's Split

First of all, microbes often make new microbes without having sex. One very common form of asexual reproduction is called *binary fission*.

As you might expect from the word *fission*, in this process a cell gets bigger, makes copies of everything inside it, and then just … splits. Presto, there are two new identical cells, each one equipped with everything it needs to go on living.

Most unicellular organisms reproduce by fission. Binary fission goes by many other names: mitosis, cell division, cell duplication, or replication.

Fission and the Bacterial Growth Curve

When most people use the word *growth*, they usually mean something is getting bigger. When microbiologists use it, very often they mean an increase in cell number. Of course bacterial growth also includes an increase in size (that is, cell mass), because the bugs must duplicate all their inner structures, ribosomes, DNA, cell walls and all, before they can divide in two.

Thus when microbiologists speak of the bacterial growth curve, they mean the orderly process of very rapid cell growth that they observe in their labs. If everything is to their liking, bacteria in laboratories can double their numbers at regular intervals, called generation time.

When bacteria are growing in a closed system like a Petri dish, microbiologists observe four phases in the bacterial growth curve:

 Wee Warnings

Don't expect to see the bacterial growth curve anywhere but in a lab. In the real world, bacteria must scramble to make a living like everybody else, competing with their fellow creatures for nutrients and other crucial resources. This tends to put brakes on the bugs' ability to increase their size and their numbers, if it doesn't kill them off altogether.

- ◆ Lag phase
- ◆ *Exponential phase*, sometimes called the *logarithmic* or log phase

- ◆ Stationary phase

- ◆ Death phase

def•i•ni•tion

An **exponent,** as you probably know, is the teeny number placed above and to the right of a number that indicates how many times that number must be multiplied by itself to give a product: $2^4 = 2 \times 2 \times 2 \times 2 = 16$. The exponent in that equation is 4. A **logarithm** is an alternative way of expressing an exponent for the number 10. Log 5 = $10^5 = 10 \times 10 \times 10 \times 10 \times 10 = 100,000$.

The lag phase begins when the bugs are placed on the growth medium. A medium is a kind of tool or carrier; in microbiology it's a mixture of nutrients that make it possible to nurture microbes in the lab.

During this phase, bacteria appear to just sit there, but in reality they are getting ready to grow in number. They are making enzymes, proteins, and RNA, in general boosting their metabolic activity in preparation for the explosion of activity that is about to occur. No fission yet.

When fission begins, the bugs have entered the exponential phase, the logarithmic, or log phase. In ordinary conversation, when people speak of "exponential growth," they usually mean simply that the growth is very fast. They are expressing the idea that the larger a number gets, the faster it grows. Exponential growth in bacteria is measured by generation time, which is the amount of time a population of bacteria takes to double in number. Depending on conditions and bacterial species, the new generation can appear in 20 minutes or 20 hours.

The exponential phase can be maintained by giving the bugs a steady supply of fresh growth medium or by transferring some of the culture to a new supply of fresh medium.

In the stationary phase, the population number stabilizes. Cell division and cell death are in balance; that is, fission creates new bacteria as fast as old bacteria die off. Bacteria begin to die because the nutrient supply is running out or because toxins they give off have built up to a lethal level.

The death phase begins when fission stops or when the death of cells outpaces replication. No new cells are created; there is only death. Note that the bacteria die off at a rate that is exponential, too, just like the fast-growth log phase.

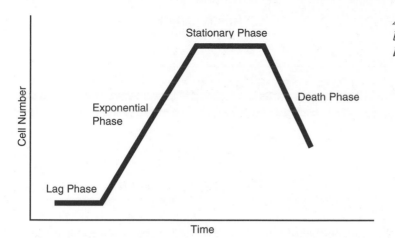

A schematic graph of the bacterial growth curve in the lab.

Swapping Genes Without Sex

Microbes do have ways of exchanging genetic material with others of their species to make a new organism with traits of each parent organism, although these activities are not exactly sex. They are conjugation, transformation, and transduction.

Conjugation only occurs when a bacterium contains mobilization elements that permit transfer. In conjugation, one bacterium sends pieces of genetic information to the other through contact between the two cells. The receiving bacterium then incorporates some of this DNA into its chromosome (or plasmid). The receiving bug is now a hybrid that combines its original DNA with some from the donor bacterium. But the donor bacterium is not changed at all.

Another method for DNA acquisition is called *transformation:* in transformation, the bug takes up bits of DNA from outside sources and incorporates it into its own DNA. These sources can include related species, plasmids, viral genomes, or even genome bits that dead bacteria have left lying around. Don't be too grossed out by that. Your own precious DNA is stuffed with castoff genetic material from other organisms, especially viruses.

Conjugation

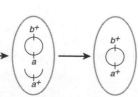

| Donor Bacterium | Recipient Bacterium | DNA Transfer from Donor to Recipient | Recipient Takes Up New DNA | Recombination |

In conjugation, one bug transfers DNA to another via direct contact between their cells. The recipient now has some donor DNA incorporated into its own genetic material, but the donor is unchanged.

In transformation, a recipient bacterium takes up DNA from one of several sources. Here cell lysis of a donor bacterium has made DNA available for incorporation by others.

Transformation

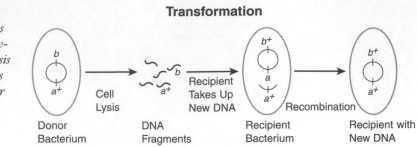

The third method is called *transduction*. There are two sorts of transduction. One occurs in nature, and the other occurs in the lab.

In natural transduction, foreign genes get into bacteria via one of the transport "vehicles" we discussed in Chapter 2 and 4: namely the viruses called bacteriophages, which infect bacteria. As you'll recall, they charge in, hijack the bacterium's machinery for copying its own DNA, which forces the bacterium to copy the virus's genetic material. These copies are turned into virions.

A few of the new virions may contain no DNA at all, but sometimes they also contain bits of the host bacterium's DNA as well as the virus's. When these hybrid viruses infect other bacteria, they can transfer the original bacterium's DNA to the new bacterium's genome, where it behaves just like any other DNA. It will be incorporated into the new bug's genome and passed on to its descendants.

In the lab, molecular biologists employ transduction for a number of purposes. These days one of the most common reasons is to make model organisms for studying disease, which we have discussed before.

In transduction, a phage infecting the donor bacterium picks up some of its DNA. The phage triggers cell lysis in the donor, and then the phage carrying some bug DNA infects another bacterium. The phage injects its hybrid DNA into the recipient bug, which incorporates the donor bug's DNA into its own.

Transduction

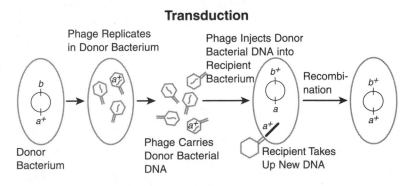

In addition to fission, fungi also can reproduce asexually in a number of ways. The yeast that makes bread dough rise, *Saccharomyces cerevisiae*, buds off cells that become new yeast. Part of the budding yeast protrudes, grows, and eventually detaches itself from the mother yeast.

Little Did You Know

You can think of yeast as making the ultimate maternal sacrifice. Every time a mother yeast cell buds off a daughter yeast cell, the process creates a bud scar on the mother cell. You can figure out how many times a cell has budded by looking at it through a microscope and counting the bud scars. When the mother cell is entirely covered with these scars, she dies.

Other fungi, and some algae, commonly produce spores. A fungal spore is a one-celled reproductive structure that is often compared to a plant seed, although it doesn't pack a lot of food inside it like a seed. Fungi produce spores in huge numbers, and they are carried to new locales by wind, water, and passing animals.

Coming Together

Some unicellular microbes can also reproduce sexually. In some protozoa with flagella, for example, the two parent cells simply fuse, even their DNA-containing nuclei. This new cell—called a *zygote*—then splits, undergoes binary fission, and forms two new flagellates.

Some ciliates make babies via conjugation. The two parent cells attach to each other but do not fuse. At the attachment point, little holes called pores develop and the two parents exchange reproductive nuclei. An exchanged nucleus fuses with the resident nucleus to form a zygote with genetic material from both parents. This zygote can undergo fission several times and produce many offspring that way.

Fungi engage in a number of forms of sexual reproduction, including fusion and some spore production.

Among the many things that microbiologists don't know yet is exactly how much sex goes on among microbes. It is generally believed that most microbes reproduce asexually only, but every now and again a microbe will give microbiologists a peek around the curtain and they get a surprise.

Take, for example, *Candida albicans*, which causes an annoying vaginal infection, the main symptom of which is an infuriating itch. This yeast infection can be transmitted between sex partners, but *C. albicans* is also a normal, harmless resident of the human vagina. It can grow wildly under certain metabolic conditions—for example, if antibiotics taken to cure a sinus infection also kill off competing organisms that usually keep the yeast in check.

Scientists used to believe this yeast reproduced only asexually. Not long ago they learned that it possesses some genes similar to those the related yeast *Saccharomyces cerevisiae* employs for sexual reproduction. With that hint in hand, further research showed that *C. albicans* does mate occasionally.

The Least You Need to Know

◆ All microbes must take in carbon, oxygen, hydrogen, calcium, potassium, and other nutrients, along with tiny amounts of trace elements.

◆ Microbes "eat" in various ways, including diffusion or active transport across the cell membrane.

◆ Like other life-forms, microbes need certain conditions to grow—such as the right amount of oxygen, the right temperature, and the right pH.

◆ Microbes often reproduce by simply dividing in two, a process called *fission*.

◆ Under laboratory conditions, members of bacterial colonies increase their numbers in a four-stage process known as the bacterial growth curve.

◆ In addition to fission, many microbes can also engage in other forms of making offspring, including sex.

Part 3

Life on Earth—and on Man, Woman, and Child

This section explores one of the most intriguing microbial traits: how sociable they are. Microbes are not solitary. They live in communities. In fact, some of them *are* communities to themselves: microbes inside microbes inside microbes. We'll tell you how the immune system works to fend microbes off—and how the wily microbes have found ways to evade it. We'll tell you about ways of preventing disease and treating it. And—perhaps most fascinating of all—we'll tell you all about those nine out of ten of your cells that aren't you at all, they're microbes.

Chapter 9

The Social Life of Microbes

In This Chapter

- Intimacy, microbe style
- The community in the slime
- Microbes are codependent, too
- One-sided microbial relationships
- The enemy within
- The origins of organelles

You are not a single organism. You are a community of trillions of organisms, home to many more microbes than there are cells in your body. Most are bacteria, but some are fungi and protozoa. Many do you good. Some are essential; you couldn't live without them. Only a few do you harm.

Just about all microbes live in communities. This networking is a fact of life and has been for well over a billion years. It is called *symbiosis*. Just like humans, microbes have symbiotic relationships with each other and with other organisms.

Symbiosis and Endosymbiosis: Living Together

Symbiosis literally means "living together" and describes an intimate association between two or more different kinds of organisms. Often this association is so very intimate that one organism actually lives inside the other, a relationship known as endosymbiosis.

Scientists have traditionally classified symbiotic relationships as …

Wee Warnings

The partners in symbiotic networking are called sym-bionts. A few authorities reserve the term symbiont to mean only the smaller partner in a symbiosis. They call the larger partner the host. We will use the term host, too, but when we say *symbiont,* we usually mean either partner.

- ◆ Mutual, when they are beneficial to both part-ners.

- ◆ Commensal, when they benefit only one of the symbionts.

- ◆ Parasitical, when one exploits or injures (and sometimes kills) another.

The Feeling is Mutual

The mutual relationship is just like a marriage between two organisms. There is give and take and the partners both benefit or suffer.

The quintessential form of mutualism is biofilm, layers of slime that are really layers of microbes. Biofilm depends on quorum sensing, communication and behavior coordination among a group of microbes, to function. We talked a bit about that in Chapter 2 when we discussed how vibrios produce light.

Biofilm: Studying Slime

In the past, microbiologists have concentrated on studying single microbes that they would grow into colonies in their labs. These free-swimming bugs are known as planktonic bacteria. Now, scientists understand that nearly all microbes are not free-swimming. Instead, they live in communities and conduct their lives quite differently in the real world than they do in the lab. Thus, more and more scientists are examining microbial communities.

A biofilm is a microbial community. It is a layer of slime that requires water and attaches itself to rocks, wood, cement, plastic, metal, glass, to say nothing of plant tissue and animal flesh. According to the experts, there is no such thing as a biofilm-proof surface.

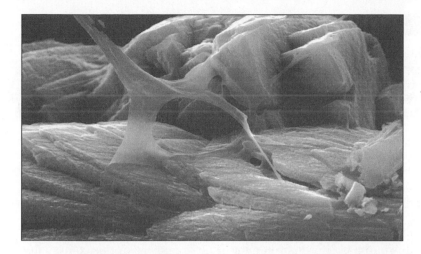

Sheets of sticky biofilm, in actuality diverse microbe communities, form on rocks and just about any other surface where water is available.

(Courtesy of NASA/JSC.)

Living in a community like this protects individual members. It safeguards them against rain, drying out, sunlight, wind, unpleasant temperature changes, poisonous substances produced by their neighbors, and host defenses that might kill them. Biofilm microbes share food and even genes; the community is the perfect place for horizontal gene transfer between different kinds of microbes, and a lot of it goes on there. Microbes in biofilm can even form microenvironments within the film that provide them with just the temperature and pH they like best.

The community begins with planktonic bacteria that attach to a surface. These bugs exude a sticky polysaccharide that acts like glue, anchoring the organisms to the surface. This glue is called the extracellular polymeric substance and often contains protein components as well.

These bacteria are soon joined by many other bacteria, but often all the other kinds of microbes as well and even larger organisms, depending on environmental conditions. A white mat of microbes that lies under a 600-foot Antarctic ice shelf is a centimeter thick and topped by clusters of clams.

A mature biofilm generates a tough matrix of sugars and proteins that protects the colony. Mature biofilm is that thin layer of green goo that collects on your fish tank, but it can also form clumps, ridges, streamers, and even structures that look like tiny mushrooms. It is often riddled with channels that let food in and flush waste products out. Microbial mats that form on water surfaces are one type of biofilm and almost always contain lots of photosynthesizing microbes that can create their own energy.

Eventually planktonic bacteria swim away from the colony. They seek out their fortunes on other surfaces, beginning a new biofilm.

Biofilms can be helpful, but they can also do harm. Biofilms burn holes in concrete and corrode computer chips. They eat away at and plug up water pipes. They pour toxins into their surroundings.

Behold the pink slime, for example. It floats on top of green pools with a pH like battery acid, 1,400 feet down in a California iron mine. The organisms in this community eat iron, fix nitrogen and carbon, and also produce sulfuric acid, which dissolves iron and other metals. These dissolved metals flow out of the mine into streams, making the mine the largest single source of toxic metals in the United States.

Scientists doing genome sequencing on this community found DNA from at least five organisms, four of them brand new to science. They included protists, fungi, bacteria, and archaeons.

Quorum Sensing: Communication, Coordination, and Competition

Biofilms couldn't form without quorum sensing. They need a certain concentration of the bacterial chemicals known as pheromones or autoinducers to estimate cell density, to attract additional residents, and to signal other microbes that it's time to turn genes on and off. When only a few bugs are around, the signals aren't strong enough.

One well-studied example of a chemical signal in gram-negative bacteria is N-acyl homoserine lactones (AHL). As you'll recall, biofilm bacteria possess a receptor that can detect AHL. Receiving the signal causes different sets of microbial genes to turn on, including genes that produce more inducers.

There are many signaling molecules, an unknown number of which are still to be discovered. These constitute different "languages," so not all microbes can understand the signals of other microbes. However, there is evidence for some communication between different species; it's called cross-talk. Some bugs use cross-talk as a weapon, detecting competitive species and killing them.

More Mutualism

Fungi infecting plant roots help eight out of ten plants get minerals and water from the soil. The fungi are known collectively as *mycorrhizae* (singular: *mycorrhiza*).

This mutually beneficial symbiotic association goes back more than 450 million years to the beginnings of land plants. In fact, scientists now believe that plants would never have made it on the land, at that time a dry and barren place, without the help of friendly fungi in their roots to scarf up essential minerals and scarce water and pass them along to their hosts in exchange for sugar snacks. Mycorrhizae can also protect plants from pathogens and toxic wastes, and, by conserving water and preventing erosion, improve the soil.

Rhizobia: Bugs in the Beans

Bacteria also have entered into mutualism with plants, most notably the bacteria that live among roots of legumes, which are plants in the pea/bean family. That mutualism is based on nitrogen, an essential element for making proteins and nucleic acids.

All life needs nitrogen. Air is nearly 80 percent nitrogen, but only a few organisms, such as bacteria in the genus *Rhizobium*, can use gaseous nitrogen from the air. So where do other organisms get the nitrogen they need?

Animals get nitrogen from plants, either from eating plants or eating animals that ate plants. In order for plants to use nitrogen, it must be "fixed" or integrated into compounds like ammonia.

Legumes have rhizobia in their roots. These bugs supply legumes with fixed nitrogen from the air, using a great deal of energy to process it into the form the plants can use. In return, the legumes supply the bugs with bed, board, and oxygen. Other soil bacteria—for example, *Azotobacter*—can fix nitrogen by themselves, but rhizobia must be living in legume roots to do it.

Rhizobia moved into legume roots much more recently than fungi moved into plants. This particular mutualism occurred only toward the close of the Dinosaur Age 70 million years ago. Yet it employed a tactic similar to that of mycorrhizae many eons before. Scientists have discovered that the bacteria learned how to colonize their green hosts by exploiting some of the same plant genes that the fungi had exploited already.

These plant/microbe relationships may seem like bliss, but as we all know, even good relationships have their bad moments. Scientists are beginning to recognize that terms like mutualism and commensalism can be fluid and variable and do not always imply harmony. Rhizobium is enshrined in most microbiology books as the classic example of mutualism. But in this relationship of so-called equals, one symbiont, the plant, turns out to be more equal than the other. When the bacteria cut back on nitrogen production, their soybean hosts punish them by shutting down the bugs' supply of food and oxygen.

Ruminating on Mutualism in Animals

Several bacteria are known to have cozy inside relationships with insects, and there are probably many more such mutualisms that have yet to be unearthed. Bacteria from the genus *Buchnera*, for instance, provide essential amino acids to aphids. Our favorite, though, is bacteria with the lovely name *Wigglesworthia*, which supply B vitamins to tsetse flies.

On a larger scale, ruminate for a moment about ruminants. This animal group includes not only cattle but other large vegetarians, among them goats, camels, and even giraffes. Ruminants have a four-compartment stomach that employs a polyglot biofilm containing a team of dozens of species of bacteria, archaea, protozoa, and even fungi. They make enzymes the ruminants lack. These microbial enzymes break down otherwise indigestible cellulose, the mainstay of the ruminant diet. The nutrients are then released not only to supply energy, but also to be turned into vitamins.

This digestive process is simply biofilm formation. The cow chews greenery, breaking it up. Free-swimming bugs in the rumen (the first stomach) glue themselves to cellulose in the crushed plants with their special sticky polysaccharides. They also generate enzymes that break up the cellulose into sugars and other chemicals.

These sugars interest other bacteria, including anaerobes. The anaerobes turn the cellulose's sugar into organic acids, which in other circumstances would retard bacterial growth. But by this time other microbes have joined the party. They include the archaea known as methanogens, which munch on an increasing buffet of organic acids. So the bacterial population continues to grow, and the archaea turn the acids the anaerobes make into the gas methane.

The ruminant then shamelessly betrays its microbial benefactors by using its last two stomach compartments to digest them, too. Another illustration that mutualism does not necessarily describe a relationship wholly benign for all the symbionts all the time.

> **Little Did You Know**
>
> The indispensable biofilm in ruminants' stomachs may be a big contributor to global warming. Methane is the foremost product of their biofilm metabolism, but it is also a major greenhouse gas that traps heat in Earth's atmosphere. Thanks partly to Cows & Co., the atmospheric methane supply keeps growing. Up to a quarter of that gas is estimated to come from ruminants—most of it apparently from burping, not farting.

Commensalism: Is It Real?

Officially commensalism describes a kind of symbiosis in which one partner benefits and the other is neither benefited nor harmed. That may define some relationships in larger organisms—for example, a dung beetle that collects tortoise feces for storage and later dining pleasure. But it's not clear that the official definition applies very often in microbiology.

In microbiology, scientists often use *commensalism* to mean something like: "A relationship in which Microbe X appears to reside in Organism Y as the normal state of affairs." For example, when textbooks speak of commensals in the human gut, they almost always mean the microbes normally found there, even though some of those microbes are known to be beneficial and others are at least potentially harmful.

The idea that commensalism is a symbiotic state where one of the partners (usually the host) is unaffected may mean simply that a potentially pathogenic microbe only goes to work when the environment changes, for example, when a host's immune system is damaged. We have become thoroughly familiar with this dangerous phenomenon in AIDS, where patients often get so-called opportunistic infections, sometimes deadly ones, from microbes that couldn't gain a foothold in a healthy person.

Or the idea that commensalism means that one of the symbionts is unaffected may simply reflect the current state of microbiologists' knowledge. They may know the bug is there, but they don't yet know exactly what it's up to. In this case, defining a specific symbiosis as commensal may really mean, "This relationship is probably either mutualism or parasitic, but we haven't figured out the subtleties yet."

In Chapter 4, we talked about the infuriating vaginal infection due to the yeastlike fungus *Candida albicans*. *C. albicans* is passed as a normal, harmless commensal from mother to newborn. It is a lifelong resident in the mouths and GI tracts of most of us and rarely causes trouble.

Or take the protozoan *Giardia intestinalis*, a.k.a. *Giardia lamblia* or *duodenalis*. The Dutch scientist van Leeuwenhoek discovered this flagellate in his own feces; he described them in an 1681 letter as "a-moving very prettily…" with the help of flagella the Dutch scientist said were like "sundry little paws." *Giardia* is found in many animals and up to one in three people.

Until late in the 1960s, this ubiquitous creature was thought to be a harmless commensal because *Giardia* infections are usually symptom-free. Actually, *Giardia* turns out to be the cause of giardiasis, one of the most common kinds of diarrhea. The waterborne protozoa are highly contagious between humans and can also be acquired from animals. Wrapped in their protective cysts, *Giardia* survives easily not just in lakes and streams but in swimming pools, hot tubs, and even municipal water supplies.

Parasites: The Unwanted Dinner Guest

The ex-commensal *Giardia* is now routinely referred to as a parasite. The term parasite is said to be derived from a Greek word referring to a guest who comes to dinner and stays. As we said earlier, a parasitical relationship is one where one symbiont exploits or injures (and sometimes kills) the other.

Wee Warnings _____

What the microbe does to the host may well depend on circumstances. The fungus *Colletotrichum magna* causes disease in watermelons and squash. But when researchers infect tomatoes with *C. magna*, they get big, yummy, disease-resistant tomatoes. In this case, the nature of the symbiosis between fungus and plant depends on the plant's genes.

There are many organisms classified as parasites. A lot is known about them because they cause disease and so have been heavily studied. Strictly speaking, we suppose it might be said that all pathogens are parasites by definition, since they get something out of the relationship but make the host ill in the process, sometimes fatally ill. In medicine, however, the term parasite seems to be reserved for eukaryotes, not prokaryotic pathogens.

As it happens, parasitic diseases are some of the most important in the world. Parasitic nematodes like roundworm and hookworm infect nearly a billion people. And even when we are talking only about microbes, the parasite burden is staggering. Half a billion people, one in twelve of our Earth's human inhabitants, suffer from malaria, which causes 3,000 deaths a day. Malaria is a protozoan infection transmitted through mosquito bites, and we have a good deal to say about it in Chapter 16.

Although most microbial parasites of interest to people are protozoa, fungi can be parasites, too. Fruit growers worry about *gymnosporangium*, a group of fungus species with an intriguing tree-based lifestyle that damages apples, pears, and other fruit. The fungus grows into striking orange balls in junipers, and then it forms and releases spores that drift on the wind until they can infect nearby fruit trees. Those fungi then form new spores that are released to infect local junipers. This use of alternate hosts doesn't hurt the junipers much, but it reduces fruit production substantially. Orchardists "cure" the infection by getting rid of nearby junipers, depriving the fungus of one of the hosts it must have to survive.

The fungal parasites of most human interest are those that infect skin and nails to cause ringworm, jock itch, and other dermal annoyances.

We talk more about microbial parasite diseases of people, especially in Chapter 11. But for a moment let's consider parasites that infect other creatures.

What is the most frequent source of bacterial infection on the planet? Probably *Wolbachia*. This bacterium does not cause human disease, but *Wolbachia* is so common because it infects a significant proportion of a million species of insects (at least one out of five and in some populations three out of four). It also infects spiders, nematodes, and other life-forms without backbones. Some of these infections appear to be mutualism; the host depends on the bug for survival. But when *Wolbachia* behaves like a parasite, its effects are so various and bizarre that it has scientists agog.

Parasites are good at making their hosts behave in ways that will help the parasites produce more parasites. *Wolbachia* is absolutely brilliant at manipulating its hosts in these unprecedented ways.

- Commands hosts' reproductive systems to produce more eggs (which *Wolbachia* can infect) and little or no sperm (which are too small for infection)

- Increases its own numbers by making it hard for uninfected host females to reproduce

- Produces a substance that makes the eggs of infected host females more viable than those of uninfected females

- Forces host wasps to produce only parthenogenetic females, which reproduce with no male help

◆ Modifies its male wasp host's hormones so as to force him to produce eggs

◆ Kills all host eggs that would produce males and allows hatched host females to feed on the remains

◆ Permanently changes the proportion of females to males in a host species by selectively destroying males

◆ Hastens the process of forming new host species because closely related hosts infected by different *Wolbachia* strains don't mate

Taking a lesson from *Wolbachia*'s ability to manipulate its hosts, scientists are hoping to do a little manipulating of their own. They would like to turn *Wolbachia* into a weapon against its host when the host is an agent of disease in humans or the plants and animals humans depend on.

Wolbachia infection races through a host population like wildfire. Perhaps, researchers say, they could turn the bacterium into a vector for quickly transporting detrimental genes into a host population. Those genes could be designed to cripple the host's ability to create problems or even kill it off.

Researchers already have reported success with attacking *Wolbachia* as a way to get rid of a host that depends on it. Filariasis, more commonly known as elephantiasis, is a debilitating tropical disease in which people's limbs swell up horribly. It results from infection by threadlike worms which are in turn infected by a number of *Wolbachia* species.

Treatments that attack the worms directly are not notably successful. But in a pilot study of some 70 filariasis patients, eight weeks of treatment with an antibiotic not only killed off the bacterium *Wolbachia*, it also got rid of the worms that apparently cannot live without their infection. Researchers are hoping an antibiotic approach will work with other worm infections, too, especially a tropical disease commonly known as river blindness, a major cause of blindness.

The Astonishing Origin of Organelles

We come now to the most special case of endosymbiosis; it's special because no development has been more important for life on Earth. The eukaryotic cell, it turns out, is actually a symbiotic coalition of prokaryotes. At least some of their organelles used to be microbes that traded their talents to other microbes in exchange for room, board, and safety.

The Energy Organelles: Mitochondria and Plastids

Eukaryotic cells have many kinds of organelles, which are literally little organs. Organelles permit different parts of the cell to specialize in different jobs, so many scientists think it was organelles that made the eukaryotic lifestyle possible. Ribosomes, where proteins are made, are organelles. So is the nucleus that houses the eukaryote genome and defines eukaryote life. So is the lysosome, one of the cell's garbage disposal units, which cleans up by digesting invading bacteria.

Here we focus on just a few organelles that are particularly important for microbiology because it is now clear that these organelles originated from microbes—bacteria, to be precise. This is the endosymbiotic theory of organelle origins. It holds that a number of cell organelles (most notably *mitochondria* and plastids) originated when a prokaryote "ate" another prokaryote, but, instead of digesting this microbial meal, put it to work for the benefit of both.

def•i•ni•tion

Mitochondria are specialized cell structures that are the principal source of energy in almost all eukaryotic cells. These organelles are often called the power plants of the cell, and a cell can contain thousands of them.

By compartmentalizing the cell's factories for producing energy, this union made it possible for multicellular organisms to proliferate. And that proliferation led to increasingly complex life-forms, including the one that has begun to figure out how all this happened. (That's our own, *Homo sapiens*, in case you were wondering.)

Scientists think that mitochondria originated about the same time as eukaryotes, more than 1.5 billion years ago. All existing mitochondria seem to be descended from a single event in which an aerobic bacterium something like today's α–proteobacterium moved into a host that was something like Archaea. The bacterium probably resembled today's Rickettsia, a group of submicroscopic bugs that live inside the cells of other creatures. Rocky Mountain spotted fever is caused by one with the redundant name *Rickettsia rickettsii*, which lives in ticks.

Scientists aren't sure whether the original host was already a eukaryote. Perhaps not, since all mitochondria studied so far, from any species, possess a few genes from the original bacterial immigrant. That genetic similarity is also the reason they think this event happened only once.

Chloroplasts, remember, are derived from plastids. Plastids came into existence only a bit later than mitochondria, when a cyanobacterium moved into a mitochondria-carrying eukaryote. This, too, is believed to have happened only once. But the ability to photosynthesize was exceptionally useful, and it was transferred laterally to other eukaryotes many times. In 1905, their early discoverer, Konstantin Mereschkowsky, called plastids "little green slaves."

Scientists are still arguing about the origins of some other organelles, but here's how they think it happened for two kinds of organelles that make the energy cells run on: mitochondria, which live in both animal and plant cells, and chloroplasts, derived from plastids and found only in plants.

Many prokaryotes can make energy anaerobically in their cytoplasm. They don't need oxygen for the chemical reaction that turns one molecule of the simple sugar glucose into four molecules of adenosine triphosphate (ATP), which stores and transports energy.

The anaerobic process (called glycolysis) isn't terribly efficient. Once free oxygen is available, however, a far more cost-effective energy process becomes possible. This different sort of chemical reaction (called the Krebs cycle or the citric acid cycle) transforms a single glucose molecule into two dozen or more ATP molecules.

In the beginning there was hardly any atmospheric oxygen on Earth, so life had to get along without it. In fact, you'll remember, oxygen was poison to creatures that existed then, and still is to most anaerobes.

But life changes and adapts. About 3 billion years ago, evolution created photosynthetic bacteria, probably cyanobacteria. Photosynthesis produces energy, but you'll recall that it also produces free oxygen as a byproduct. The more these early photosynthesizers turned light into metabolic energy, the more oxygen they released into the atmosphere.

About a billion and a half years ago, enough oxygen had built up so that some bacteria adopted the aerobic life. They began to use oxygen to produce much more ATP energy with much less food.

Eventually some eukaryotes slurped up these aerobes, providing them with food and shelter in exchange for the energy they could produce. The result was eukaryote cells containing the remnants of two kinds of bacteria that had turned themselves into two organelles, mitochondria and plastids.

There are several lines of evidence to back up the endosymbiotic origin theory of mitochondria and plastids. Here are some:

♦ Mitochondria and chloroplasts contain their own DNA, and it is more similar to the DNA found in prokaryotes than it is to the eukaryote genome.

◆ The organelles are enclosed by cell membranes, and the inner membrane is like a prokaryote membrane.

◆ These organelles make new copies of themselves via binary fission, just as prokaryotes do.

◆ Chloroplast biochemistry is similar to the biochemistry of the cyanobacteria they probably came from.

> **Little Did You Know**
>
> Human mitochondria have retained only 13 protein-coding genes, but that doesn't mean those genes are unimportant. Mitochondrial genes have been linked to several human diseases. They also appear to be involved in the aging process.

◆ Nuclear DNA in eukaryotes has some sequences that are quite similar to sequences in chloroplasts, and are believed to have moved from the organelle to the nucleus via horizontal gene transfer.

Note the last point especially. The ex-microbes that became DNA-containing organelles gave up a lot of their genes. Some they simply lost. Others moved into the host's nuclear genome, where they remain today. Depending on the host species, mitochondria retain between 3 and 67 structural genes that produce proteins, and chloroplasts possess between 50 and 200.

Secondary Endosymbiosis

Remember diatoms from Chapter 3? Diatoms and some other algae, it turns out, are like nested Russian dolls: a microbe inside another microbe inside a third microbe. The relationship is known as secondary endosymbiosis.

The littlest microbe in the diatom stack was a cyanobacterium scarfed up by a red alga in the very distant past. Descendants of the bug became chloroplasts, and eventually a diatom engulfed both organisms and commandeered many of their genes via horizontal transfer. Recent research has confirmed these ancient events by showing that the diatom genome contains gene sequences that are very much like sequences from free-living cyanobacteria and red algae.

Scientists suspect that other examples of secondary endosymbiosis await exploration. In fact, some believe that secondary endosymbiosis that transfers photosynthetic ability has occurred many times and has played an essential role in the development of many kinds of eukaryotes.

The Origin of the Endosymbiotic Theory of Organelle Origins

The endosymbiotic theory of the origins of mitochondria and plastids is now almost universally accepted, but not so long ago it was controversial as can be. First proposed more than a century ago to account for the origin of chloroplasts, the theory was resurrected several times. Most life scientists laughed at such a flaky notion.

They kept right on laughing for a while after biologist Lynn Margulis published *Symbiosis in Cell Evolution* in 1981. She argued that a number of eukaryote organelles (not just mitochondria and chloroplasts but also cilia and flagella) had originated as prokaryotes, and that endosymbiosis (rather than genetic mutation) is the major engine of evolution.

Eventually the evidence, especially the growing DNA evidence, persuaded her colleagues that she was right about mitochondria and plastids. Even after a billion years inside other life-forms, the energy organelles hang on to traces of their bacterial origins. In fact, these bacterialike characteristics helped convince scientists that endosymbiosis theory really did explain how these organelles came to be.

Scientists largely remain skeptical about some of Margulis's other ideas. A microbial origin for cilia and flagella seems doubtful to most, for example. These organelles contain no DNA that would reveal where they came from. But while researchers may continue to think that mutations chiefly drive evolution, today most concede that microbial symbionts have also been essential to the dazzling complexity of life on Earth.

The Least You Need to Know

- Symbiosis is an intimate association between two or more different kinds of organisms.

- Mutualism is a symbiotic relationship in which the symbionts benefit each other.

- Quorum sensing, communication and behavior coordination in a community of microbes, is essential to biofilm formation.

- In parasitism, one symbiont exploits (and sometimes kills) the other; many disease organisms (especially eukaryotes) are called parasites.

- Endosymbiosis is a special kind of symbiosis in which one symbiont lives inside the other.

- In secondary endosymbiosis, both an endosymbiont and its microbial host are engulfed by a third microbe.

The Immune System: Life in the Combat Zone

In This Chapter

- ◆ The immune system, distinguishing self from nonself
- ◆ The innate immune system and the adaptive immune system
- ◆ Cell-mediated immunity
- ◆ Humoral immunity and antibodies
- ◆ The immune system remembers
- ◆ Immune system diseases

Organisms are constantly at war with each other, fighting for food and space and other resources they need to live. Microbes battle among themselves, but they also attack more complex organisms, which have in turn developed sophisticated ways of protecting themselves.

This chapter describes some of these details of immune systems, including the intricate methods organisms use to identify and defend against outside forces, such as microbes. It is necessarily only a brief account, with emphasis on the systems that serve humanity so well. In later chapters we discuss strategies for combating disease by exploiting immune system mechanisms—vaccine creation, for example.

The Struggle Against Disease

An organism's immune system has a single overarching purpose: to recognize components that belong to the organism itself and to tell those components apart from components that are foreign and should be destroyed. In immunology, this is called distinguishing self from nonself.

Tiny Tips

You will not be surprised to learn that some microbes have gotten good at evading surveillance and attack by immune systems. That is one way they make people sick. We discuss some of those microbial strategies in later chapters.

The mechanism for recognizing "self" is proteins of the major histocompatibility complex (MHC) produced by the organism's own genes. Almost all vertebrate cells possess MHC proteins on their surfaces. The human MHC proteins are called human leukocyte antigens (HLA).

The mechanism for recognizing "nonself" is a foreign antigen. An antigen is any substance that can trigger a response from the immune system. Antigens include the majority of macromolecules, nearly all proteins, and many polysaccharides as well.

For discussion purposes, immunity is often divided into two systems: innate and adaptive. The innate immune system, as the name implies, is the system that organisms are born with. But to call it a "system" is something of a misnomer because, in fact, it's a miscellany, a heterogeneous collection of strategies for battling invading organisms. Innate immunity consists of general methods of coping with a variety of invaders.

By contrast, in the adaptive (also called acquired) immune system, the body responds to new pathogens by "learning" their characteristics and developing ways of fighting them off. Adaptive immunity is very specific; it develops a defense against a single kind of invader—sometimes for life.

The adaptive immune system is the crown jewel of the immune system. It employs structures of dazzling inventiveness that first appeared about 470 million years ago, with the debut of vertebrates with jaws. All jawed vertebrates possess an adaptive immune system, so that includes all mammals—and, of course, us. The adaptive immune system is able to deal with an almost unlimited array of potential pathogens.

These two systems, innate and adaptive, entwine in many ways. Cells active in the innate system also do their work as part of adaptive immunity. It is not always possible to classify a particular protective mechanism cleanly as innate or acquired.

The Innate Immune System

Most organisms larger than a single cell—plants as well as the tiniest animals—possess inborn ways of fighting off pathogens and other potential invaders. These mechanisms mostly act quickly to stave off danger. They are not designed for specific invaders, but are rather general defensive mechanisms that work against a variety of invaders.

Barrier Methods

The first defense tactic is to present barriers to would-be invaders. Microbes, for example, usually possess cell walls, sometimes quite thick, that are difficult for a pathogen to penetrate. Humans and other animals use barriers to keep pathogens out, too. Some of the barriers include skin, expulsion, mucous, pH, and toxic secretions.

One of the most important barriers is skin. Microbes normally can't penetrate skin unless there's a breach in the barrier: a cut or scrape that exposes flesh and makes it available for colonization.

We can banish pathogens simply by ejecting them. Urine, tears, and saliva wash them away. In the lungs, little hairlike structures called cilia sweep them up and out, and coughing and sneezing helps us rid our airway of the pathogens.

Mucous in our airways and intestinal tract entraps and immobilizes microbes. Other creatures (amphibians and fish) are coated on the outside with this sticky microbe-trapping stuff.

The acidic pH of the skin's surface repels invaders on the outside, and inside, the stomach is so acidic that only a few specially adapted microbes can survive there.

Body fluids contain enzymes that smash through cell walls of gram-positive bacteria, causing them to burst. Virtually every human body fluid uses defensive enzymes. They are found in saliva, tears, that stuff in your nose, breast milk, even sweat and semen.

In addition to these defenses, plants and animals also possess antimicrobial peptides called defensins. They can kill both gram-positive and gram-negative bacteria, plus other microbes: fungi, protozoa, even viruses.

Calling On Defenders

We also marshal microbes to fight microbes. The trillions of microorganisms that normally live in and on us are a formidable defense system, for us and for other animals. They produce toxins and enzymes that they use to outcompete (and attack) invading microbes. We discuss this normal flora in more detail in Chapter 13.

We borrow our native microbes' talents in our defense, but we also make our own defenders when we distinguish between self and nonself. Nonself of course means invading microbes, but also cancer cells, cells from tissue transplants, and some toxins.

Cells of the immune system do this job by recognizing certain molecules often found on foreign cells, including pathogens, but not on host cells. Examples include molecules found in microbial flagella and the peptidoglycan in bacterial cell walls. Chief among these host immune system cells are the leukocytes, white blood cells.

There are several different kinds of leukocytes that defend vertebrates against infection. As you might expect, they carry out several different kinds of jobs. The most common leukocytes are the neutrophils, which account for more than half of the white blood cells. Neutrophils are critical to the immune system's response. A neutrophil (also called polymorphonuclear leukocyte, PMN, and "poly") resides in the blood and is a kind of phagocyte. *Phagocyte* is from Greek words meaning "cell eating," and that is exactly what phagocytes do: surround, swallow up, and digest other cells, mostly dead cells and microbes. This engulfing behavior is a form of endocytosis. This method of fighting invaders by eating them appeared quite early in the history of Earth life.

Phagocytes affix themselves to microbes and cover them with proteins. They then consume the microbes and chop them up with enzymes. Finally, phagocytes plaster their cell surfaces with pieces of the microbe; these are then known as antigen-presenting cells. The antigens are attached to MHC proteins. Putting them on exhibit like this permits them to be recognized as antigens by other cells of the immune system.

Neutrophils make heavy use of defensins for exterminating the pathogens they slurp up. Neutrophils have short lives, ranging from a few hours to a few days. Exceptionally mobile, the "first responders" of the immune system's white cells race to the scene of an infection, especially infection by bacteria and fungi. The pus that forms in infected wounds is mostly carcasses of neutrophils that have done their work and expired.

Phagocytes are drawn to infection sites by chemicals given off by invading microbes, by immune system molecules, and by damaged cells. Neutrophils are drawn to the infection site by macrophages. The macrophage is a large white blood cell that is another kind of phagocyte. *Macrophage* is from Greek words meaning "big eater." Although macrophages at the infection site help draw in neutrophils, a few hours later many more macrophages turn up.

Long-lived, macrophages reside in tissues all over the body. They lurk particularly in places vulnerable to invading organisms such as the lungs and the spleen. They also possess cell surface receptors that can identify, and stick to, some molecules on the surfaces of bacterial cells. Macrophages are scavengers, recognizing, engulfing, and digesting pathogens and other microbes, especially those that live inside cells (and also other debris). Macrophages also release proteins known as cytokines that are a major communication method between immune system cells. Cytokines attract other immune system workers, such as neutrophils.

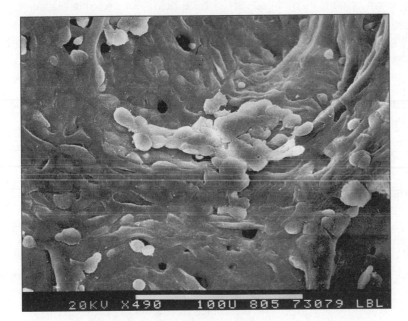

Scanning electron micrograph of a clump of bacteria-fighting macrophages clinging to the wall of a lung.

(Courtesy of the Lawrence-Berkeley National Laboratory.)

Microbes have learned how to resist phagocytosis. Some gram-positive bacteria coat themselves with slimy capsules that effectively hide their telltale cell-surface molecules from phagocytes on patrol. Some bugs, *Mycobacterium tuberculosis* for one, manage to survive phagocytosis and carry on within macrophages. Viruses, too: macrophages can themselves be infected by HIV and carry this cache of replicating deadly viruses throughout the body.

Tiny Tips

Although macrophages are part of the innate immune system, they also work for the acquired immune system.

Part 3: Life on Earth—and on Man, Woman, and Child

Despite their main role as body defenders, macrophages can also be involved in causing diseases; for example, diseases of the immune system itself. Macrophages play a major part in developing the plaques that are characteristic of atherosclerosis. In the process of attacking infected cells, they can damage surrounding uninfected ones.

The reason neutrophils and macrophages get to an infection site so quickly is the complement system. Among the antimicrobial proteins drawn to infection is complement. Complement is a cascade of dozens of enzymes that augment—complement—the actions of other immune system components. Complement covers microbes with molecules that mark them for swallowing up by phagocytes; it shatters them, enhances inflammation, and boosts its own production.

The qualities that make complement so effective also make it potentially dangerous to other tissue, so there are mechanisms that can turn it off quickly once it's dealt with pathogens. One is instability; complement proteins fall apart quickly if they can't find targets. Another is host cells, which contain inhibitors that turn off the complement cascade. These mechanisms are not always effective, which is why researchers suspect that complement figures in autoimmune diseases such as arthritis and multiple sclerosis.

Complement and other immune system molecules also alter the permeability of blood vessels, permitting fluids, white cells, and proteins to pour into the infection point. This is inflammation. Infection by pathogens almost always leads to inflammation, a distinctive combination of redness, heat, swelling, and pain at the site. Inflammation is triggered by chemical signals from damaged tissue and from the invading microbes themselves. These signals also attract neutrophils and other leukocytes that attack and engulf the invaders.

Although inflammation can be therapeutic because it is a magnet for immune system cells that will attack the pathogen, it also can be dangerous. When molecules that trigger inflammation pour into the bloodstream carrying the infection with them, the result can be fatal septic shock, a widespread inflammatory process that can lead to massive organ failure. Inflammation in the wrong place at the wrong time can also cause chronic disorders such as cardiovascular disease and asthma.

And microbes have figured out ways to exploit the inflammatory response to their benefit. When *Salmonella* triggers inflammation, it attracts macrophages, which the bug then hijacks for transportation to other parts of the body.

The Adaptive (Acquired) Immune System

The phagocytes of the innate system are very good at their jobs, but they are not able to identify all invaders. For that, what's needed is the other immune system, the adaptive (or acquired) immune system. The adaptive system responds to foreign antigens more slowly than the innate system, but it is far more specific, ordering up custom-designed proteins called antibodies to fight the invaders. And unlike the innate system, the adaptive system has memory. When it encounters the outsider again, it remembers its previous defense and can mount it much more quickly.

Lymphocytes

The adaptive immune system makes use of another kind of white cell: the lymphocytes, T and B cells. Most parts of the immune system are pretty amazing, but these lymphocytes may be the most amazing structures of all.

They can produce an immune response designed specifically to battle almost any conceivable antigen. That's because each individual lymphocyte carries at least one unique receptor that can bind one particular kind of antigen. Together the body's trillions of lymphocytes possess a mammoth inventory of receptor possibilities equipped to deal with almost any antigen that comes by. Up to half of our lymphocytes move through blood circulation, but the others travel in a different circulatory system, the lymphatic system. The lymphatic system is the central hub of the immune system in most animals.

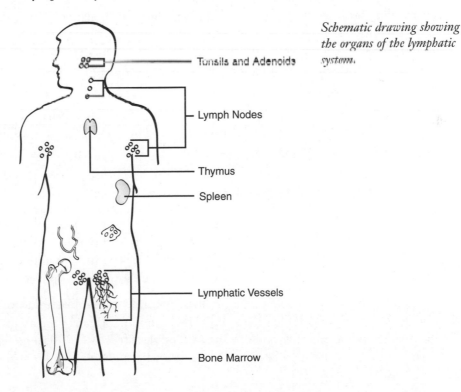

Schematic drawing showing the organs of the lymphatic system.

Tonsils and Adenoids

Lymph Nodes

Thymus

Spleen

Lymphatic Vessels

Bone Marrow

The human lymphatic system consists of organs (bone marrow, spleen, thymus, tonsils, adenoids), lymph nodes (in the neck, armpits, groin), and vessels that carry a quart or two of lymph. The basis for lymph is plasma (clear fluid) that has leaked from the blood circulatory system into surrounding tissue, where it is collected by lymph system vessels and eventually returned to the circulatory system.

Like all blood cells, lymphocytes begin as undifferentiated cells in the bone marrow. Some remain in bone marrow until they mature and are released; these are the B lymphocytes. Some move on to the thymus for maturation; they are released as T lymphocytes. Most lymphocytes, about 80 percent, are T cells. On release, both B and T cells collect in lymphoid organs, awaiting antigens that will activate them.

The lymphocytes are the heart and soul of the adaptive immune system because they can modify their behavior to, well, adapt to novel kinds of antigens. They do this in two ways: cell-mediated immunity (carried out by T cells) and humoral immunity (carried out by B cells).

Cell-Mediated Immunity

Cell-mediated immunity works mostly with two kinds of T cells. Cytotoxic T cells especially combat viruses (and also bacteria that infect cells) by identifying hijacked cells carrying foreign antigens and persuading them to commit suicide. Cytotoxic T cells learn to recognize antigens that are combined with MHC proteins and are shown to them by antigen-presenting macrophages and other immune system cells.

Helper T cells trigger macrophages to go forth and seek invaders. They also produce interleukins, a group of cytokines that lead to rapid creation of new T and B cells. Depletion of T cells is the major problem posed by HIV infection. A third kind, regulatory T cells, oversee this system.

Humoral Immunity and Antibodies

The humoral immunity system is based on antibodies (immunoglobulins) produced by B cells. Each B cell carries many copies of a single antibody on its surface, and antibodies also enter the circulation. Antibodies stick to foreign antigens and act as a kind of label declaring that the antigen should be destroyed by phagocytes.

Antibodies must be capable of enormous structural variation in order to be able to recognize the essentially limitless number of foreign antigens they might encounter. So an antibody is another one of nature's modular structures. It is a molecule made of four peptide chains—two heavy chains and two light chains—and is shaped like a Y.

The ends of these chains, the tips of the Y, are short, variable amino acid sequences. In any one antibody, the variable sequences are identical on both tips. These variable regions have different shapes. They are what give antibodies their ability to stick to just about any antigen, marking it for destruction. There are billions of potential antigens, so there are billions of potential antibodies.

The rest of the Y is called the constant region. Remember that antibodies are also called immunoglobulins, abbreviated Ig. They differ in where they operate in the body and how they trigger reactions from the innate immune system. There are five different kinds of constant regions:

Tiny Tips

Antibodies are big enough to block viruses and prevent them from entering cells.

- IgG accounts for about 80 percent of serum antibodies in humans and is the only one that can cross the placenta and confer some immunity on fetuses and newborns. It attacks both bacteria and viruses.

- IgM accounts for about 10 percent of antibodies and makes bacteria clump together, resulting in easier consumption by phagocytes.

- IgA accounts for about 5 percent of antibodies; it shields mucosa from infection and is also present in breast milk, tears, and saliva.

- IgD accounts for less than 1 percent of antibodies; it is plentiful on B cells and serves as a trigger for antibody production.

- IgE accounts for only a tiny proportion of antibodies but is responsible for a good deal of misery because it is central to allergic reactions.

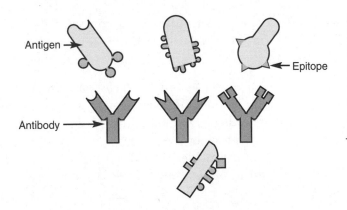

Antigen

Epitope

Antibody

Schematic drawing showing how antibodies capture and bind foreign antigens, marking them for destruction. The Y-shaped antibodies recognize and fit into unique molecular structures on the antigen surface called epitopes.

Antibody genes are limited in number. So how do antibodies achieve their incredible diversity? Because antibody construction is modular and antibodies are made of different peptide segments that can be mixed and matched.

During B cell development, individual fragments of DNA that make parts of the different light and heavy chains are shuffled around randomly via genetic recombination. Thus each B cell is randomly programmed to make one light chain and one heavy chain different from the other B cells.

This recombination procedure alone can generate many millions of possible combinations of antibody fragments. Other mechanisms increase the chance for nearly endless variety. As it turns out, antibody-making DNA is particularly prone to mutation, further increasing the possibilities for brand-new random amino acid combinations. Mistakes during the recombination process also lead to new permutations. The result is that each of us has the potential for making billions of unique antibodies that can do battle with the billions of potential foreign antigens.

The Immune System's Memory

One of the most extraordinary innovations of the adaptive immune system is that it employs memory. It can recognize a previous pathogen when it occurs again and quickly recall how to fight it. This is why there are many infectious diseases you can get only once.

The memory system works this way. B cells carrying an antibody that recognizes and binds a specific antigen then go on to divide, creating additional copies of themselves and their successful antibody. These daughter cells are called plasma and memory cells.

Tiny Tips

Immunological memory is also the underlying principle of vaccination, where administration of a harmless or crippled version of a pathogen is enough to stimulate a long-term immune response without making the host actively ill.

Plasma cells tend to die fairly rapidly, but memory cells persist and continue to divide. T cells also make memory cells that continue to recognize a foreign antigen long after the original infection.

Secondary immune responses are not just faster than the original; they also tend to be much stronger, resulting in antibodies that bind an antigen much better. This increased strength is the result of mutation in the variable region DNA fragment and therefore comes about by chance.

Immune System Development

A fully functioning human immune system takes a while to develop. Fetuses possess mature T and B cells, but they don't do anything until a fetus encounters a foreign antigen. During a normal pregnancy, that shouldn't happen. Pregnant women transfer some immune system molecules across the placenta into the fetus.

After birth, the antibody system develops gradually. Some antibodies show up soon after birth, but others may not show up until the child is a toddler. That gradualism also is true of other immune system components. Thus small children are not fully protected against all of life's antigens. New mothers help give babies a healthy start when they breast-feed, transferring antibodies and many other parts of her own immune system as part of the meal. Breast-fed babies are much less likely to develop intestinal and respiratory infections and allergies than are babies fed cows' milk or human milk substitutes.

Diseases of the Immune System

The immune system itself can cause disease. Autoimmune disease results from failure of the body's ways of recognizing "self." When one of these mechanisms goes awry, the body interprets its own cells as foreign, and the immune system attacks them. The results can be catastrophic.

Among these diseases, which number in the dozens and mostly affect women, are multiple sclerosis, psoriasis, type 1 diabetes, lupus, and Crohn's disease. Sometimes these immune system failures come about because of infection by microbes that cross-react with self antigens, making them appear like outsiders to the immune system.

Allergies, which affect a great many of us and can even cause death, also result from an immune system breakdown or, rather, an immune system overreaction to pollen, dust, and other harmless proteins. In allergies, IgE provokes an inflammatory response that can lead to annoying consequences like a runny nose or deadly ones like anaphylactic shock.

An absent or weak immune system is also dangerous. Everyone has heard of the Bubble Boy—a child forced to live walled off from the germ-filled world because he was completely unable to mount defenses against pathogens. Children with severe combined immunodeficiency, known as SCID, have inherited defects in B cells and T cells. There are other genetic immune system disorders as well, such as defective neutrophils.

As we have already mentioned, an immune system weakened by illness such as HIV infection or even just old age is susceptible to opportunistic organisms, microbes that normally cause no trouble but can gain the upper hand in a host with damaged defenses. Malnutrition can also damage immunologic mechanisms.

Finally, transplant rejection: this sometimes-fatal event is a result not of a defective immune system, but of an immune system that is just doing its job. Transplanted organs that are not a very close immunological match to a patient's tissues are likely to be recognized by the immune system as nonself and attacked.

The Least You Need to Know

- The "immune system" is actually several different but interacting methods of fighting off invading organisms. It recognizes antigens, substances (usually macromolecules) that can trigger an immune system response.

- Unique proteins of the major histocompatibility complex (MHC) are present on almost all vertebrate cells and help each organism distinguish self from nonself by recognizing its own cells.

- The innate immune system is a collection of nonspecific methods of fighting off invading organisms and relies, in part, on barriers such as skin and the mucosa.

- The adaptive (acquired) immune system is capable of custom-designed methods of fighting off specific pathogens.

- The adaptive immune system's chief weapon is antibodies (immunoglobulins), which can be constructed in billions of configurations to combat any one of billions of potential foreign antigens.

- The immune system remembers past infections and can quickly mount a strong response when a pathogen reappears.

Dodging Disease by Preventing Pathogens

In This Chapter

- ◆ Keeping clear of pathogens
- ◆ Killing germs with sterilization
- ◆ Reducing microbes with disinfectants
- ◆ Preventing infection with antiseptics
- ◆ Protecting humans one shot at a time

Humanity has made remarkable strides learning how to treat and cure diseases after they occur, and we take up that topic in Chapter 12. Even better than curing disease is preventing it in the first place. Prevention is the subject of this chapter.

Staying Away from Pathogens

The very best way to avoid disease, of course, is to avoid the pathogens that cause it. That's especially true with some microbes, notably viruses, where trying to treat an infection after it has been acquired is not particularly satisfactory.

So, for example, infection with HIV (which eventually causes AIDS) can be prevented by avoiding exposure to bodily fluids that contain the virus. That means avoiding unprotected sexual contact and contaminated blood. Blood banks are vigilant about checking their blood supplies, discarding blood infected with HIV and other viruses like the ones that cause hepatitis B and C.

Some viral infections can be prevented by shielding people from the animals and insects that can transmit them. Human rabies is controlled by destroying rabid dogs and other rabid animals and vaccinating healthy animals before they can be exposed. Yellow fever no longer exists in the United States because the mosquitoes that carry it are controlled by insecticides and habitat controls like draining the swamps where they live.

Sanitary measures such as sewage and water treatment have played an enormous role in shielding people from infections they can get from drinking water contaminated by feces from infected people (medical folks call this "the fecal-oral route"). This has pre-vented diseases caused by all kinds of microbes: viruses (hepatitis A), bacteria (cholera), and protozoans (amebiasis).

The other major preventive strategy is to stimulate the human immune system to defend itself preemptively, before pathogens get a chance to launch infection. The chief weapon is vaccination. Vaccination has worked against many infectious microbes. We'll be talking about mobilizing the immune system for self-defense at the end of this chapter.

Getting Rid of Microbes—or at Least Keeping Them in Check

There are a number of approaches to controlling and eliminating microbes, especially pathogens. Among them are sterilization, disinfectants, and antiseptics, as well as a num-ber of drugs for treating infection: antibiotics (effective against bacteria) and other antis—antifungals, antivirals, and antiparasitic drugs. We'll define all these later in the chapter.

There are many possible ways to put a microbe out of commission, but a lot of them will damage the host as well. The challenge is to kill or cripple a harmful microbe without injuring the host. These strategies could all be lumped under the heading of "antimicrobials." Inevitably, these methods kill harmless microbes, too. They reduce the total microbial population, not just the population of pathogens.

The chief mechanisms for controlling and eliminating microbes are …

♦ Disrupting microbial membranes.

♦ Damaging microbial nucleic acids.

◆ Damaging microbial proteins (which means messing up their normal 3-D structure).

◆ Thwarting microbial metabolism.

Microbes differ in their susceptibility to these methods. Warmer temperatures often render organisms more vulnerable than colder ones. Susceptibility can depend on what growth phase a microbe is in. Actively growing microorganisms are more vulnerable to antimicrobials than those in a dormant stage—which is why spores tend to be especially resistant to control methods. Biofilms protect their microbial residents, so biofilmed surfaces must be cleaned before sterilization or disinfection.

Sterilization, disinfectants, and antiseptics, aimed mostly at bacteria, are strategies that tend to be less expensive, widely available, and usually aimed at getting rid of microbes in general before they can infect, rather than targeting specific infections after the fact. These tactics are aimed mostly at bacteria.

The term *antibiotic* is generally reserved for a manufactured drug designed for therapy of particular bacterial infections. Administered to humans or animals, an antibiotic usually kills or cripples bacteria only, not other microbes. Originally, it meant an antibacterial agent derived from other organisms, for example, penicillin from bread mold. Now it also includes the few synthetic antibacterials for treating diseases that people have invented, such as sulfa drugs.

Sterilization

Sterilization is always *bactericidal*. The object of sterilization is to exterminate all forms of life completely—no exceptions. A sterile surface by definition doesn't harbor any kind of microbes or spores that could germinate into microbes.

def•i•ni•tion

Stuff that kills bacteria is called **bactericidal** (or sometimes germicidal). But if a method only stymies bacterial growth and reproduction, it's known as bacteriostatic. Antibiotics can be either bactericidal or bacteriostatic. Disinfectants and antiseptics are usually bactericidal, although at low concentrations some can be bacteriostatic only. Microbe populations can often rebound after bacteriostatic treatment. Still, in many situations, even reducing the pathogen population can be therapeutic, since it buys time to permit a host's natural defense systems to swing into action.

Sterilization with Heat

The chief form of physical sterilization is heat. In fact, heat is probably the cheapest and most widely used approach to getting rid of microbes. Heat denatures their proteins, disrupts their cell membranes, and destroys their nucleic acids.

Among the kinds of heat employed are …

◆ Incineration at a temperature of at least 500°C (932°F), which vaporizes organic matter most efficiently, leaving only a little ashy waste behind.

◆ Boiling at a temperature of at least 100°C (212°F) for at least 30 minutes, which kills all microbes but may not kill bacterial endospores.

◆ Autoclaving (or pressure cooking, which is steam under pressure) at a temperature of at least 121°C (249.8°F) for at least 15 minutes at a pressure of at least 15 lbs/sq. in.

◆ Dry heat or an oven at a temperature of at least 160°C (320°F) for at least 2 hours.

The disadvantage of heat is that it can't be used on objects that would melt or otherwise be destroyed. Also, some methods are not entirely reliable at destroying spores or prions.

Incineration was used to destroy microbes long before anybody knew there were microbes. Human corpses (and their possessions), as well as animal corpses, have been burned during epidemics. They still are. In the 1990s, officials ordered millions of European cattle to be slaughtered and burned in attempts to prevent the spread of prion diseases to other hosts, including people. More recently, millions of chickens have gone to their deaths in order to slow the transmission of avian flu and reduce the possibility that it will turn into people flu.

Little Did You Know

Perhaps the most famous historical example of public health incineration was used against the Black Death—bubonic plague—in fourteenth-century Europe. In attempts to keep the plague from spreading, the corpses of millions of people (and their clothes and household goods) were burned.

Boiling can be convenient, but the boiling treatment that will kill microbes can leave endospores intact. One way of handling that problem, although it takes time, is to subject the material to three bouts of boiling and cooling spaced a day apart. The idea is that allowing the boiled material to cool off will force the spores to germinate, rendering them vulnerable to the effects of subsequent boiling.

There are many kinds of autoclaves, but they are all basically steel containers with pipes running in and out. Autoclaves are standard lab equipment. They generate steam under pressure, which is an efficient and relatively easy way of sterilizing any equipment or other item that can tolerate 121°C (249.8°F) for at least 15 minutes under a pressure of 15 lbs/sq. inch. The World Health Organization advises that an ordinary home pressure cooker can serve as a reasonable substitute for an autoclave. Autoclaving kills bacteria, fungi, viruses, and even spores, but not prions.

An autoclave produces moist heat, and moist heat does a faster job than dry heat. But dry heat can be a lot more convenient, since ovens are far more common than autoclaves or pressure cookers. The usual advice is that the material should be baked at 160°C (320°F) or more for at least two hours, but 171° (339.8°F) for one hour is the equivalent, and so is 121°C (249.8°F) for 16 hours. You can see that moist heat with pressure makes a big difference in sterilization times.

Tiny Tips

The opposite of heat is cold, and cold is also an effective way to control microbes. It doesn't usually kill them, but cold can slow their growth and reproduction, and severe cold (−20°C [−4°F] or less) can stop it, often without killing the microbe. This technique is central to food storage, which we discuss in Chapter 17. But it's also important for research. An ultra-cold freezer (−80°C [−112°F]) is an excellent way to store microbes for long periods, which is why many microbiology labs possess one.

Sterilization with Radiation

Another form of physical sterilization is radiation, which works by demolishing an organism's nucleic acids. Sterilization devices use several kinds of radiation: X-rays, gamma rays, infrared, microwave, and other radiation. It can kill spores as well as organisms, but not necessarily viruses. This technology can be expensive and it is often necessary to shield equipment operators from radiation damage.

One widely used form of radiation sterilization is often natural and free: ultraviolet radiation (that's sunlight). You are probably too young to remember that, before there were disposables, diapers were made of cloth, and freshly laundered diapers flapping on clotheslines in the breezy sterilizing sunshine were a common sight. There are also less natural versions of UV light: germicidal lamps.

UV light can be destructive, especially to human tissue and plastics, and it bleaches fabrics. It's useful for sterilizing surfaces but cannot penetrate glass, water, or layers of dirt very well.

Sterilization with Chemicals

It has been really hard to come up with chemicals that are excellent at getting rid of microbes while posing no risk for other organisms. As a result, some antimicrobial chemical weapons now in use have been selected because they are not very toxic, even though they are only so-so at dealing with pathogens. Prions are particularly good at resisting most attempts at sterilization.

Some gaseous chemicals used for sterilization include ethylene oxide, chlorine, and ozone. Ethylene oxide is a penetrating gas much valued for sterilizing because it can get through packing materials and plastic wraps. It can also sterilize laboratory and hospital equipment that can't be heated. The gas kills microbes and their spores by attacking their proteins.

Using ethylene oxide requires special equipment in the form of a sterilizer that looks a bit like an autoclave. The gas's effectiveness varies with temperature and humidity, so the sterilizer controls those factors as well as the concentration of gas. The gas is so toxic that sterilized material must be aired out after treatment in order to dissipate it.

Chlorine is a ubiquitous element in living things and in the oceans. It is familiar, in combination with the element sodium, as a component of table salt. But it is also a poisonous gas. Dissolved in water, chlorine gas is the most frequently used method for purifying public water supplies in the United States. It kills algae and bacteria.

Ordinary household chlorine bleach, a 5.25 percent solution of sodium hypochlorite, is usually diluted before use. It sterilizes most effectively when it is allowed to work for at least 20 minutes. In addition to killing microbes, chlorine bleach can also destroy some kinds of spores.

Ozone, a form of oxygen made of three atoms instead of "regular" oxygen's two, has a bad reputation as a pollutant and can be poison to life of all kinds. But it is also an effective alternative to chlorine for sterilizing municipal water supplies. The food industry employs ozone to kill yeast, mold, and spores on fruits, vegetables, and processing surfaces. A big advantage to ozone is that it can be produced on demand from atmospheric oxygen with electricity, so it doesn't need to be stored or transported.

Sterilization with Filters

Filters don't kill microbes, but they can keep them away efficiently, which, in many laboratories, is just as good. For example, filters made of diatomaceous earth or synthetics can block particles as small as 0.2 μm, pores fine enough to bar all microbes except viruses.

Filters such as surgical masks and the cotton stoppers on lab culture dishes are so good at blocking microbes that they permit entry of sterile air and only sterile air.

Biological safety cabinets using high-efficiency particulate air (HEPA) filters stream microbe-free air across the cabinet opening, protecting research workers in the lab (and the lab itself) from the microbes under study in the cabinet. This filter does its job so well that researchers can even work with dangerous disease organisms.

Disinfectants

Disinfectants mostly seek to kill microbes on nonliving surfaces (such as kitchen counters or surgical instruments) using physical or chemical methods. Disinfectants can sometimes sterilize, but the chief point of disinfection is to reduce the number of microbes. A related term is sanitization, where microbe populations are reduced sufficiently for public health purposes.

Most disinfectants are chemical. Disinfectants usually work against a wide range of microorganisms—not just bacteria, but viruses and microbial eukaryotes, too—although a few tend to be resistant (and spores often are). Disinfectants may be dangerous to other life forms as well, and should be handled with care.

Some common examples of disinfectants that almost everyone has used include these:

+ Chlorine (often in laundry or swimming pools), which we discussed in the section on sterilization

+ Alcohol (ethanol or isopropyl alcohol, which is also used on skin)

+ Hydrogen peroxide (its oxygen works against anaerobes but it can be toxic to normal cells)

+ Surfactants like soap and detergent

Alcohol

Alcohols penetrate best when mixed in a solution of about 80 percent alcohol and 20 percent water. While alcohols work on bacteria, fungi, and even some viruses, they cannot kill spores. They can also cause tissue damage when used on wounds. However, they do evaporate easily so they are commonly used as skin disinfectants before an injection, for example.

Hydrogen Peroxide

Hydrogen peroxide is not really a disinfectant or an antiseptic. It is good at killing pathogenic anaerobes, especially in deep wounds, but that is because it poisons them with oxygen. For this purpose, it is used in low concentrations, typically about 3 percent in solution with water.

Surfactants

Surfactants dissolve in water but can also dissolve lipids. As you'll recall, lipids don't usually dissolve in water, but surfactants make them soluble. You employ surfactants several times a day (we hope!) in the form of soap.

Soaps are derived from fats and have a pH that is alkaline (greater than seven). Many bacteria dislike high pH, so soap works by harming them. Soaps also work by simply rinsing microbes off surfaces (and your skin) and sending them away down the drain. Recent research has shown that hand-washing with plain old soap can reduce by half the two most common causes of death in childhood: diarrhea and pneumonia.

Detergents are surfactants that have been synthesized. Some are cations (positively charged) and others are anions (negatively charged). The cationic detergents kill bacteria more effectively. These cationic detergents, specifically quaternary ammonium compounds, are present in many household cleaners.

But there are many other disinfectants. The following are a few more examples.

Heavy Metals

Heavy metals like arsenic, silver, and mercury have long been used as disinfectants. Copper sulfate, for example, can kill pond algae, and selenium kills fungi. They work by inactivating microbial proteins. Heavy metals tend to be bacteriostatic rather than bactericidal, but they are very toxic to other organisms, including people.

Phenol and Phenolics

Phenolics are a broad class of chemical compounds found in everything from estrogens to capsaicin, which is the stuff that makes chili peppers hot. But only some phenolics can serve as disinfectants. They work by denaturing microbial proteins and disrupting their cell membranes.

Phenolics smell unpleasant and irritate skin, but they can deal with some hard-to-kill microbes, and a single application is active for a long time. Phenol, the simplest of the group, has been employed as a disinfectant during surgery since 1867. Phenolics continue to be used in hospitals and labs today, and household cleaning products such as Lysol also contain them.

In fact, households are now a significant source of phenolics because they have been incorporated into so many everyday products. There are antimicrobial soaps, mouthwashes, toothpaste, and cosmetics that contain phenolics and other antimicrobial chemicals. These chemicals even have been added to plastic food containers and children's toys.

Have We Made the World Too Clean?

Have antimicrobials made us too clean for our own good? Constant exposure to antimicrobial household products is something that worries specialists who have been trying to understand recent worldwide increases in asthma and allergies, both immune system disorders.

Some research suggests that the ubiquity of antimicrobial products may be robbing youngsters' immune systems of a chance to learn how to deal with the real world—that is, a world that is normally dirty and filled with microbes. The result, they argue, is that immune systems lash out at harmless targets like pollen, causing the symptoms of asthma and allergies. This idea, called the hygiene hypothesis, is by no means fully accepted yet. But scientists around the world are studying the possibility that too much cleanliness is creating disease while trying to prevent it.

Some experts are now concerned that phenolics (including an especially common one called triclosan) will create microbes that have learned how to resist them.

Antiseptics

Antiseptics (derived from Greek words meaning "against" and "putrefaction") are also aimed at preventing infection. These chemical methods interfere with microbial growth and reproduction and either cripple or kill them. Antiseptics are used on the outside of the body—that is, on the skin.

All antiseptics are also disinfectants, so the list includes many of the methods discussed in the previous section. But not all disinfectants are antiseptics. That's because, although they harm microbes, antiseptics by definition do not harm living tissue. Thus, antiseptics are on the whole not as successful against microbes as disinfectants are, and also tend to cost more. Alcohol is an antiseptic, and so is hexacloridine, a chlorinated phenolic.

Vaccines: Preventing Infectious Diseases

The notorious influenza pandemic of 1918 struck one out of every five people on the planet and killed 50 million of them. As we write this, public health officials are holding their collective breaths. They fear a flu pandemic at least as big as the 1918 pandemic some time in the next few years. Eventual mutations are expected to make it much simpler for deadly bird flu viruses to attack humans much more easily and spread from human to human. That is not the case at the moment, thankfully, but the existing bird viruses have killed more than half of the small number of people they have infected.

Public health officials know we are not much better prepared for a flu pandemic than our ancestors were nearly a century ago. Yes, in many parts of the world improved hygiene and other public health measures will help reduce the casualties. You probably live in one of them. But many people are not so lucky.

And there is not yet a cure or valuable treatment. We discuss antiviral drugs in Chapter 12, but suffice to say that they are not the answer to population-wide flu prevention.

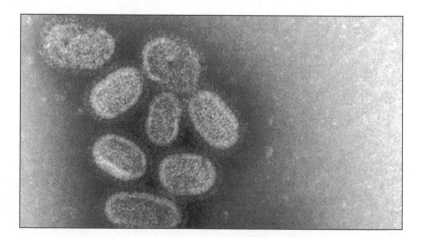

Virions of the 1918 flu, recreated by scientists in 2005. Researchers recreated this microbe in order to study what made it so contagious and deadly and to try to develop new vaccines and treatments to combat it.

(Courtesy of the CDC, Cynthia Goldsmith.)

The Vaccine Success Story

Vaccines can prevent flu, but the virus evolves faster than manufacturers can keep up with. So, while the vaccine you get today offers some protection against flu, it is really designed to combat last year's virus.

Yet vaccines are a great success story in the fight against infectious disease. Most of us don't even know anyone who has contracted one of the several near-vanished diseases that were terrifying only a generation or two ago. They include smallpox of course, which we discussed in Chapter 4. Also on the list: pertussis, diphtheria, meningitis, measles, polio, rubella, and typhoid.

But the list of extant infectious diseases for which there is no vaccine is even longer. At the top of everybody's list is HIV/AIDS. Not far behind are malaria and tuberculosis. Among others on this list are …

> **def•i•ni•tion**
>
> A **vaccine** is a preparation of particular antigens that activates the immune system without producing disease. The point is to persuade the immune system to mount a defense—to make antibodies—against specific pathogens, a defense that will be ready if infection does occur. Vaccines are often administered by injection, but some are given by mouth.

- Several sexually transmitted infections besides HIV (herpes simplex, *Chlamydia*, gonorrhea, syphilis).

- Respiratory infections (respiratory syncytial virus, parainfluenza, group A streptococci).

- Gut infections (*Salmonella*, *Shigella*, *E. coli*, *H. pylori*).

- Infections transmitted by vectors (dengue, Lyme disease, hantaviruses, schistosomiasis, leishmaniasis).

- Nosocomial (hospital-based) infections (*S. aureus*, *Pseudomonas*, *Enterococcus*, *Candida*).

- Miscellaneous (hepatitis C and E, cytomegalovirus, group B strep).

In some cases, promising vaccines are in development or clinical trials, and we have left most of those diseases off the list. In other diseases, there are vaccines (rabies, for instance), but for an assortment of reasons the vaccines are not widely used. There are also no vaccines for some of the diseases that biowarriors and terrorists might employ; we talk about those in Chapter 19.

The experts expect that vaccines will be able to protect against these infections eventually. But for most, eventually means several years at least. The obstacles are technical, economic, legal, and even cultural and political.

The antigens in vaccines are usually carried on a dead or crippled (attenuated) pathogen, or part of one. Sometimes (as in vaccines against diphtheria and tetanus) vaccines employ a pathogen's toxin, since it is the toxin, rather than the microbe itself, that causes disease. Some vaccines employ synthetic antigens. New approaches to vaccines are in development. For example, researchers are working on vaccines that use only a pathogen's DNA to stimulate a host immune response.

Vaccines produce an immune response in the same way that actual infection does. The host identifies the vaccine's antigens as "nonself," gets rid of them, and remembers them in case of future encounters.

Little Did You Know

Vaccination results in active immunity to a disease by stimulating antibody production. But some viral diseases can also be staved off via passive immunity: direct administration of antibodies made in another host (usually human). Passive immunity is a strategy against certain pathogens when no vaccine is available—or when people have just been exposed to a pathogen, or are about to be exposed to one, and their immune systems would not have enough time to make their own antibodies.

Vaccines are a public-health triumph that have prevented untold pain and debility and saved millions of lives. But they are not perfect. They can have side effects, some of which we talk about in the next section. Also, they do not always provide flawless protection from a disease. People who don't follow the recommended vaccination schedule may not be fully protected. Some people appear not to make the proper antibodies even after vaccination; this may be due to genetic or environmental interference with the immune system process.

Herd Immunity and the MMR Vaccine

Not everybody in a population has to be vaccinated in order for a vaccine to be extremely efficient. If a large proportion of the population has been vaccinated, pathogens will have a hard time successfully moving from person to person because the percentage of vulnerable people is small. This is called herd immunity.

The number of people in a population who must be vaccinated for herd immunity to work varies from disease to disease. Take the well-known example of rubella. Sometimes called German measles, rubella is mostly a mild disease of childhood without serious long-term consequences. But if a woman contracts rubella in the first three months of pregnancy, her fetus may suffer severe birth defects and even death.

The public health approach to preventing congenital rubella syndrome is to vaccinate children, not adults. By vaccinating children, the major source of rubella has been eliminated, pregnant women are not being infected, and the rate of birth defects due to the disease has been cut dramatically. In addition, rubella itself has just about vanished in some countries, including the United States.

Its public health benefits are undeniable, but vaccination is not without risk and side effects. These vary depending on the vaccine, but they are mostly rare and less serious than coming down with the disease.

The most common vaccine for children is MMR, which protects against regular measles, mumps, and rubella. In most cases, these viral diseases are mild, but that isn't always the case.

> **Little Did You Know**
>
> Thanks to the MMR vaccine, diseases that are very contagious and sometimes serious—now occur in less than 1 percent of the people in countries with vaccination programs.

Rubella, as you know, causes major birth defects. Measles patients frequently suffer complications that include pneumonia and encephalitis, and sometimes death. Mumps can cause male sterility. The possible side effects of vaccination may include a rash, slight fever, and body aches, but they are uncommon. In about one in every 100,000 vaccinations, a child suffers a severe allergic reaction.

Despite that comforting record, the MMR vaccine has been very controversial. It has been accused of causing autism and bowel diseases, among others. Parents in several countries have sued vaccine manufacturers, and in the United Kingdom, vaccination rates have dropped sharply.

However, large epidemiological studies in several parts of the world have shown no link between the vaccine and autism. As we write, medical authorities agree that parents do not need to be concerned that MMR will cause autism. And public health authorities point out that, as vaccination rates have gone down, measles is on the rise in the United Kingdom. Some fear an epidemic if things get to the point where the herd-immunity principle no longer applies because not enough people have been vaccinated.

Other Immunological Approaches to Pathogen Prevention

As we noted in Chapter 10, the immune system produces several different kinds of antibodies with different functions and skills, and many of them continue to work against pathogens for a long time, thanks to immunological memory. Most existing vaccines, however, stimulate only one type: IgG antibodies. Because vaccines don't usually enter the body in the same way that many pathogens do, through the mucous membranes that line the body (which is often where other kinds of antibodies are triggered), vaccine-generated protection may not last as long as natural protection.

Vaccines are not the only possibilities for enhancing a host's immune system to fight disease. Earlier we mentioned another: administering ready-made antibodies produced in other hosts as a quick tactic for fighting off a disease microbe. One advantage of this particular immunological approach is that it can work on many pathogens, not just a specific one, as vaccines do.

One group of immune-system enhancers that has generated a lot of media attention is interferons, which are cytokines. Interferons are protein components of most animal immune systems. As you might guess from the name, interferons interfere; in this case, they interfere with a virus's ability to make copies of itself in the host's cells. Antigens trigger interferon production in many different immune system cell types, a process that stimulates macrophages, which, as you'll recall, gobble up infected cells.

Therapeutic interferons are produced in large quantities with the help of bacteria engineered to contain human interferon genes, and also from mammalian cells. Interferons are used for treating hepatitis C, several cancers, and multiple sclerosis.

The Least You Need to Know

- The best way to avoid infectious disease is to avoid pathogens.

- Antimicrobial methods tend to kill or cripple microbes indiscriminately, destroying the harmless ones as well as pathogens and often pose risks to other organisms, too, including human hosts.

- Sterilization aims to exterminate all forms of life completely, with no exceptions, using heat, radiation, chemicals, and filters.

- Disinfectants are used to kill microbes on nonliving surfaces.

- Thanks to vaccination, several serious diseases that were common just a few decades ago have nearly disappeared in many places in the world.

- Vaccines, though, are still lacking against other common infections, including HIV and other sexually transmitted infections, lots of respiratory diseases, infections of the gut, and diseases that center around hospitals and other health-care facilities.

12

Dealing with Disease

In This Chapter

- Battling bacteria
- The difficult fight against viruses
- Fighting the fungus among us
- Antibiotic resistance
- Finding new antibiotics

Preventing infection by microbial pathogens is best, of course, when possible. We discussed those strategies in Chapter 11. But all too often, preventing infection isn't possible. Infectious disease happens, and that calls for treatment. There are several drug-based approaches that attempt to deal with disease by dealing with the microbe that causes it. Let's take a look.

Antibiotics: Weapons from the Microbe Wars

For billions of years, tiny organisms have engaged in an arms race, hurling toxic molecules at each other in the struggle to prosper. Nearly all of today's antibiotics are versions of weapons long wielded by microbes and fungi—especially inhabitants of the soil. Soils are enormously complex ecosystems, so it makes sense that its residents have devised ways of competing with each other successfully.

Scientists began discovering antibiotics less than a hundred years ago, in the 1920s. But they came into wide use only with the demands of World War II. Antibiotics have traditionally been molecules plucked from nature's battleground and modified a bit. Chemical synthesis of entirely human-created antibiotics yielded sulfa drugs in the 1930s, and more recently the fluoroquinolones, a group of broad-spectrum antibiotics that includes Cipro (which became famously scarce during the 2001-anthrax scare) and linezolid, which is effective against some resistant strains of *Staphylococcus*, *Streptococcus*, and *Enterococcus*.

Antibiotics from Molds: Penicillin and Friends

Antibiotic history usually begins with the time-honored (but not completely true) tale of how the Scottish scientist Alexander Fleming discovered penicillin in 1928 by noticing that there was a ring around a mold contaminating a culture of the bacterium *Staphylococcus*. Fleming deduced that the mold, *Penicillium notatum*, was making a chemical that killed the bug. He named the chemical penicillin, but decided (erroneously) that penicillin did not last long enough in the body to be therapeutic. He abandoned his studies in 1931.

In fact, the antibacterial properties of *Penicillium* had been known since at least 1896, thanks to investigations by French medical student Ernest Duchesne, whose findings were ignored. In 1938, the Australian scientist Howard Florey, working at Oxford, read about Fleming's work. He and his colleagues proceeded to extract the mold's killing chemical, established that it killed bacteria, and found a way to manufacture it on a large scale.

The penicillin derived from bread mold remains the most-used antibiotic and is still in common use today. It was the first in the exceedingly popular family of beta-lactam antibiotics. This group includes ampicillin, the cephalosporins, and many others. Cephalosporins come from a different genus of mold, *Cephalosporium*. The cephalosporins work against more pathogens than penicillin does, with little toxicity.

The beta-lactams kill bacteria by preventing the cross-links critical to the structure of peptidoglycan in bacterial cell walls. These drugs are therefore effective mostly against gram-positive bugs, although some have now been modified to work against a few gram-negative bacteria as well.

Common side effects include diarrhea and diseases like candidiasis, which are due to other microbes that overgrow when antibiotics kill off their competition. These antibiotics also cause allergic reactions in up to one in ten patients, and even occasional deaths.

Antibiotics from Man: Sulfa Drugs

But before there was penicillin, there were sulfonamides, commonly called sulfa drugs. Unlike most other antibiotics, they are human products, not derived from the microbe wars. The first was discovered in 1932 by the German chemist Gerhard Domagk, who was investigating dyes known to have antibacterial effects.

Sulfa drugs block bacterial enzymes needed to make essential folic acid. People (and animals) also need the vitamin folic acid, but we get it from food (or supplements) instead of making it, so sulfa drugs don't affect our supply. The drugs do a good job against pathogenic streptococci (especially *Streptococcus pneumoniae)* and also *E. coli*. Unfortunately, allergies to sulfa drugs are common.

> **Little Did You Know**
>
> Some protozoa also must make their own folic acid, so sulfa drugs can work against protozoan diseases, too.

Antibiotics from Bugs: The Mycins and Tetracyclines

Not all natural antibiotics come from molds. Bacteria produce compounds that attack other bacteria, too. Antibiotics in this group, first isolated from soil microbes in the U.S. lab of Selman Waksman in 1943, were curing patients by the 1950s.

Streptomyces species are the source of streptomycin, erythromycin, the tetracyclines, and others. They work against both gram-positive and gram-negative bugs by interfering with protein synthesis.

Long-term use of some mycins can damage kidneys and hearing. The tetracyclines have especially low toxicity and few side effects (except for increasing sensitivity to the sun), so they have been heavily employed to treat disease. The result is that many bacteria can now resist them, which we'll explore later in this chapter.

Bacillus species are the source of polymyxin (which damages cytoplasmic membranes in gram-negative bugs) and bacitracin (which blocks cell wall synthesis in gram-positive bugs). These drugs are relatively toxic to the host, so they are used only topically.

Combining Forces

The medical folks often use combinations of antibiotics instead of just one. This strategy can work for people who are near death or people who are infected by more than one microbe, for example a mixture of aerobes and anaerobes. Using more than one drug can also mean using lower doses of a particularly toxic one—or in cases where the docs don't know what pathogens are causing trouble. Sometimes the aim is synergy: certain drugs are more powerful when they act together, so lower doses of both are then possible. Unfortunately, the reverse is also true. Some drugs fight each other when in combination and so reduce their effectiveness.

Among the combinations known to be effective:

◆ Taking penicillin and streptomycin for treating endocarditis caused by *Enterococcus faecalis*

◆ Using two penicillins that interfere with two different stages of bacterial cell wall construction

◆ Combining two sulfa drugs that mess up different stages of folic acid manufacture

Clinicians also use combinations to slow down development of bacteria that can resist antibiotics. That is often the strategy in treating tuberculosis or some chronic infections.

Potential Harms from Antibiotics

Unwanted side effects, sometimes dangerous ones, are a possibility with any drug treatment, including antibiotics. Many antibiotics are not very toxic to the sick, which is why they are so enormously useful. But even mild side effects, like the very common diarrhea, are annoying, sometimes serious, and even fatal.

There are three main ways antibiotics can be harmful. They can mess up the host's internal ecosystem by killing off some helpful or harmless resident microbes, which reduces normal competition and permits other microbes to grow wildly and cause disease. They can be directly toxic to the host. They can combine with other drugs to become more toxic.

Each therapy has caused allergic reactions in somebody, somewhere. In addition, people differ in their genetic makeup, sometimes reacting strongly to a drug that is benign in everyone else. For the moment it is impossible to predict these reactions except in a very few cases. Scientists hope that new tools like genome sequencing will fix that problem and even deliver individually custom-tailored therapies for disease. But realization of those hopes is some distance away.

Therapies for Nonbacterial Diseases

It has been especially difficult to find drugs that work against other kinds of microbes that cause disease, like fungi and viruses. Why? Because those microbes' biochemistry is so like our own that damaging them risks damaging ourselves.

Bacteria, being prokaryote cells, present unique targets because they are so different from human cells. Fungi, however, are eukaryotes, with a metabolism similar enough to mammal cells to make it hard to find drugs that will harm them without harming their hosts. One exception is some *Streptomyces* species, the source of antifungal drugs that damage membranes.

And viruses work by seizing host cell machinery, so their metabolism is in truth their host's metabolism. The virus itself is just a dab of genetic material that presents few potential points of attack. Not surprisingly, most of the few antiviral drugs around block nucleotide synthesis. They include acyclovir (against herpesviruses) and AZT (against HIV).

Antiviral Drugs

Antivirals began to appear in the 1960s, mainly for treating herpesvirus infections. The emergence of HIV infection and AIDS in the 1980s triggered enormous interest in antivirals. There are now many in existence, most developed for HIV infection, plus herpesvirus and hepatitis.

Wee Warnings

One serious disadvantage of the current antivirals is that many tend to be toxic to the host as well as the virus. AZT is well known for causing anemia. Acyclovir causes severe nausea. Pregnant women are advised to avoid them due to a risk of birth defects.

One approach these drugs take is to try to block a virus's entry into host cells. Amantadine and rimantadine, for treating influenza, are entry blockers. Another line of attack is to create compounds that masquerade as parts of DNA and RNA, blocking efforts to synthesize genetic material. This is how acyclovir works against herpes. The HIV drug AZT works this way, too. Researchers also are exploring ways of attacking pathogenic viruses at other stages in their existence. Tamiflu, for instance, which is effective against both influenza A and B, blocks the release of viruses from the host cell.

Just like antibiotics, antivirals can be combined for more impact. People with HIV infections in particular typically must follow quite complex drug regimens that are uncomfortable, inconvenient, expensive, and hard to stick with. In addition, treatment guidelines shift frequently.

Antifungal Drugs

In the United States, government approval has been bestowed on fewer than a dozen systemic antifungal drugs—that is, drugs that affect the entire body. Most systemic antifungal drugs interfere with a component of the fungal cell membrane that humans don't possess. This component is ergosterol, which is similar to our cholesterol.

Some drugs derived from *Streptomyces*, the main example being amphotericin, cause fungal membranes to leak. Amphotericin B is effective against a wide array of mycoses, including ones that are potentially fatal. It must be administered intravenously and can result in several side effects, some of them severe. It is especially toxic to the kidneys.

Others, such as fluconazole, ketoconazole, and naftifine block synthesis of ergosterol. This group tends to cause fewer immediate side effects than amphotericin, but long-term use can lead to serious liver damage. They are also liable to drug interactions, notably with cyclosporine and some antihistamine and antiseizure drugs, as well as others.

Other antifungals use different mechanisms. One is 5-fluorocytosine, which blocks nucleic-acid synthesis. It is frequently used in combination with amphotericin B for treating cryptococcal meningitis and some other mycoses. Griseofulvin, made by a species of our old friend *Penicillium*, is used systemically for treating skin and nail fungus infections.

Many antifungals are not used systemically but rather are applied topically, to the skin and other body surfaces, to treat fungal infections that are localized. They include some drugs that perform similarly to those discussed in the previous section. Nystatin, for example, has a mechanism like amphotericin B's. It causes fungal cell membranes to leak, which kills the fungus. Nystatin is used topically to treat yeast infections, including *Candida*.

But there are a great many topical fungal treatments, often available without prescription, and they take many different chemical approaches. A lot of them are marketed without prescription for treating athlete's foot, jock itch, and other common fungal infections of skin and nails.

Resistance Is Not, Apparently, Futile

It is the certain fate of all antibiotics to be fought off eventually by the pathogens they target. Organisms cannot survive unless they learn how to evade such weapons, and microbes have had billions of years to evolve defensive strategies. This evolutionary process has been more pronounced since the invention of antibiotics. It is hardly surprising that microbes humanity thought it had defeated have popped up again—only this time armed with genes that help it laugh at the drug that would have destroyed it in the past. But the situation has been alarming public-health officials for some time.

Some microbes come already equipped to handle certain antibiotics; this is called intrinsic resistance. But bacteria often gain the ability to withstand particular antibiotics; this is acquired resistance (acquired, that is, from other bacteria that were already resistant). It has become an increasingly crucial public health problem. Some 40 percent of *Streptococcus pneumoniae* strains, responsible for everything from sinus trouble and ear infections to meningitis and, of course, pneumonia, are now resistant to both penicillin and erythromycin.

Resistance is conferred by genes that code for proteins that protect bacteria from an antibiotic. The genes that bestow resistance are sometimes simply existing mutations in only a few members of a bug species. This is evolution in action: while their vulnerable kin fall victim to a particular antibiotic, the mutants are selected for, so they survive and prosper.

These days, it is more common for bacteria to acquire resistance genes from other bacteria. These genes frequently are carried on plasmids and other mobile genetic elements, such as transposons, which can be transferred even between species. One well-known example is genes for enzymes that can break up penicillin and other beta-lactams; the enzymes are (not surprisingly) known as beta-lactamases.

Some bacteria learn how to pump the potentially deadly antibiotics out of their cells. (Some make this strategy work for getting rid of antiseptics, too.) Some are able to change an antibiotic's structure so it can no longer recognize its target or change their own structure (bacterial ribosomes, for example), making them invisible to the antibiotic. As a result, a number of pathogens now possess several genes that permit them to fend off almost all the antibiotics in common use—and to pass that collection to other microbes in a single step. This is called *multidrug resistance* (MDR). Thanks to escalating international travel and lack of uniform policies on antibiotic use, multidrug resistance is accelerating everywhere in the world.

def•i•ni•tion

MDR bacteria are bacteria that can fight off several antibiotics. They are sometimes known as "superbugs."

The more antibiotics are used, the faster resistance is transferred from one bug to another. As a result, places where many patients get antibiotics for their diseases—hospitals, day-care facilities, nursing homes—seethe with multidrug-resistant bugs. Then as patients return home, they carry MDR bugs into their communities.

You Can't Keep Them Down on the Farm

Health-care facilities are not the only places where antibiotic resistance is born. Agriculture is believed to be a significant source of it, too. Livestock, for example, are given subtherapeutic levels of antibiotics routinely to promote their growth rather than to treat a particular disease, a process that is believed to speed up bacterial exchange of resistance genes.

Take, for instance, experience with two structurally related antibiotics, avoparcin (for promoting animal growth) and vancomycin (for treating human disease). In the 1970s, avoparcin was approved for farm use in Europe but not in the United States. In the 1980s, use of vancomycin in people increased around the world, which became a problem. Bugs like enterococci and *S. aureus* had become resistant to other antibiotics. As a result, vancomycin-resistant enterococci (VRE) appeared.

A scanning electron micrograph of vancomycin-resistant enterococci (VRE). European countries reduced VRE prevalence in the general population after banning the use of avoparcin for growth promotion in farm animals.

(Courtesy of the CDC, Janice Carr.)

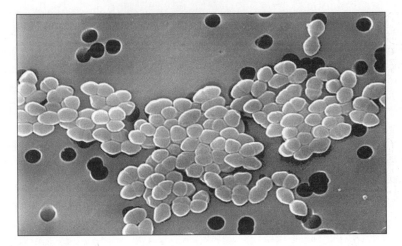

In the United States, however, VRE are rare outside hospitals except in former hospital patients. In Europe, by contrast, VRE are found in up to 12 percent of the population, including people who have never been hospitalized.

The difference, the experts believe, is that avoparcin use in livestock selected for vancomycin-resistant bacteria in the environment—in soil, in food. These resistant bacteria were not confined to hospitals, but rather were passed from person to person in communities. That avoparcin played an important role in hastening this evolutionary process was demonstrated when the European Union banned its use. The result: VRE prevalence declined.

Immersion in MRSA

Which brings us to MRSA (methicillin-resistant *Staphylococcus aureus*.) Staph are present normally in the noses of at least one in four people without consequences. Sometimes this bug causes serious disease: pneumonia plus infections of wounds and blood. MRSA are staph that are resistant to beta-lactam antibiotics (penicillin and friends).

MRSA are present in about one in a hundred people, usually patients or others connected with hospitals and other health-care facilities. But, like VRE, they have emerged outside of medical institutions as well; it is estimated that more than one in ten MRSA infections are acquired in the community. MRSA are a global problem; these superbugs are found everywhere. Risk factors include close contact with an infected person, open cuts, contaminated surfaces, and infrequent hand washing and bathing. (Hand washing should include soap and water or an alcohol-based hand sanitizer.)

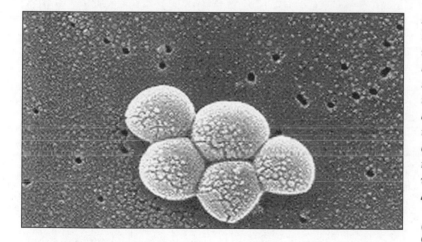

Methicillin-resistant Staphylococcus aureus *(MRSA), magnified 20,000 ×. These bugs were among the first identified that were resistant to vancomycin as well. Those clumps and bumps on the bacteria's surfaces are indications of how their cell walls have thickened beyond normal, which helps them defend against antibiotics.*

(Courtesy of the CDC, Janice Carr.)

The Difficult *Clostridium*

This nasty bug has a name that suits it perfectly: *Clostridium difficile*. Difficult indeed, and increasingly so. *C. difficile* (so named in the 1930s because it was difficult to isolate and culture) is frequently called C. diff or C-diff. The bug is a gram-positive anaerobe that forms spores. It likes soil, but it also likes the human gut; C-diff is a common commensal.

C-diff also is resistant to many antibiotics, so when a person gets antibiotic treatment that kills off many resident microbes, C-diff is a survivor. With no competition to keep its numbers down, it overgrows. If it is one of the strains that makes exceptionally powerful toxins, C-diff damages the colon, causing a severe, long-lasting, and hard-to-cure diarrheal illness and sometimes death.

C-diff spores resist heat and stomach acid, traveling to the colon. There they germinate, multiply, and produce the two cruel toxins. As is the case with other resistant bugs, C-diff disease is appearing more frequently outside hospitals, in the community and even among people who have not taken antibiotics.

Prescription heartburn medications, which are growing in popularity, appear to increase the risk of C-diff disease; reducing stomach acid is believed to make the gut somewhat more hospitable to colonization. Most C-diff, though, is still susceptible to two antibiotics: metronidazole and vancomycin.

A multitude of Clostridium difficile *in a sample of feces, magnified 1,513×. Health-care workers can spread this particularly nasty bug among hospital patients.*

(Courtesy of the CDC, Janice Carr.)

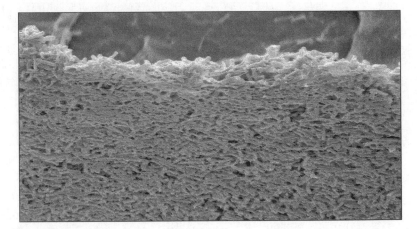

What Should We Do About Antibiotic Resistance?

One approach to slowing development of antibiotic resistance is to reduce the number of antibiotic prescriptions. Public health campaigns have emphasized in particular that antibiotics should not be prescribed for colds (because they are caused by viruses) and ailments when the culprit might be a virus rather than a bacterium (such as the minor sore throats and acute bronchitis of childhood). But many experts think much more effort should go into reducing the number of inappropriate antibiotic prescriptions.

Among the strategies for finding new ways to combat infectious disease has been increasing interest in how to take advantage of commensals, finding ways to stimulate them to fight off pathogens. To get to that point, microbiologists will have to discover much more about normal human residents and their communities. The idea is to find ways to encourage beneficials at the expense of pathogens. This is called the probiotic approach; we talk more about it in Chapter 20.

Another approach is to encourage the immune system's native ability to deal with microbial invaders, especially pathogens. This may have the added advantage of getting around the problem of antibiotic resistance. It could also expand our currently limited bag of weapons against viruses and fungi. There is even hope that beefing up the immune system may help people fight off emerging microbes or microbes spread through biowarfare, topics we take up in Chapters 17 and 19.

No one is expecting a magic bullet—an immune-system approach that will work for all people in all places against all pathogens. But there is great hope for immune-system modifications against specific pathogens or for use in particular groups of people or as additions to standard drug therapies.

How to Find New Antibiotics—and Why

Historically the usual way of finding a new antibiotic has been laborious screening of immense chemical libraries of compounds, natural and otherwise. Some argue that screening chemical libraries is approaching a dead end. But some experts think the problem with screening may simply be that libraries are not good enough or complete enough. Marine organisms have not been studied well, and 90 percent of organisms in the biosphere can't be cultured in standard ways.

One approach to developing new antibiotics is to figure out new ways to hijack the biosynthesis of antibiotics in nature so as to modify their structures with the goal of improving them. In a second approach, some researchers are trying something completely different. First, they find the most vulnerable targets in a bacterium and then design something that hits one or more of them hard. This is where microbial genome projects are going to come in handy.

Tiny Tips

New antibiotics may be the eventual outcome, but the immediate goal of research on the cell cycle is more basic. The point is not just to understand pathogenic organisms, but to understand the complete network of regulatory mechanisms that control the bacterial cell.

The process is called a rational drug design. Instead of looking at all essential genes in a bacterium and choosing one to target, some researchers are looking at genetic circuitry that controls the cell cycle, which is the pathway that coordinates cell growth and differentiation. They hope to find genes encoding proteins that control several critical functions in the cell.

Antibiotic resistance has led to new interest in phage therapy, which we mentioned in Chapter 4. One big advantage appears to be that so far no resistance has developed to phage enzymes used as experimental therapy instead of phages themselves. Even if microbes eventually develop resistance to phage enzymes, phage therapy could buy researchers decades in which to develop new antibiotics.

Decades is what they need. Antibiotic discovery is enormously difficult, and creating a practical antibiotic is a very long process. This is why trying to eliminate the unnecessary use of antibiotics—for example, prescribing them for colds—is so important.

Even proper treatment with antibiotics can create resistant strains of bugs that lurk in a treated patient well beyond the duration of the disease, for months or more. These former patients can then pass on the resistant bugs to family and friends, spreading antibiotic resistance in the population.

And even if antibiotic consumption slows down, we will still need new antibiotics. The lifetime of any antibiotic is inevitably finite, and the countdown to becoming ineffective begins the moment medicine begins to use it. We need new ways to defeat infectious disease, and we will need them forever.

The Least You Need to Know

- Antibiotics, which are used mainly against bacterial infections, are derived mostly from toxins that molds and bacteria have used to battle each other for millions of years.

- For treating an infectious disease, health-care workers often use more than one antibiotic.

- Antibiotics can cause side effects and allergies, and they usually kill off harmless and helpful microbes as well as pathogens.

- Drugs for viral and fungal diseases that are safe for humans have been hard to find because the biochemistry of these microbes is so similar to our own.

- A serious problem with antibiotics is that microbes can learn how to defend against them, sometimes creating "superbugs" that are resistant to multiple antibiotics.

- Reducing the number of antibiotic prescriptions could probably slow down development of antibiotic resistance, but it is now clear that we will always need to develop new antibiotics in order to keep up with microbes' swiftly evolving defenses.

Chapter 13

Life on Man, Woman, and Child

In This Chapter

- ◆ Human commensals, the normal flora
- ◆ Gut microbes
- ◆ Stomach microbes, especially *Helicobacter pylori*
- ◆ Skin microbes, especially herpesvirus
- ◆ Microbes of the mouth and dental biofilm
- ◆ Microbes of the urogenital tract

You are home sweet home to many microbes—mostly bacteria, but also the odd protozoan, fungus, and even archaean. This is entirely normal. In this chapter, we explain how and why.

Normal Human Flora

Scientists assume that most microbe-human associations are mutually beneficial symbioses, or at least commensal—although in many cases they do not yet know the exact nature of the association.

def•i•ni•tion

You might think that **flora** means flowers, or at least plants. But when microbiologists speak of flora, they are talking about microbes that colonize a particular place or host—usually the microbes that are normally there, not sporadic pathogens. These organisms are not plants, of course. In fact, when microbiologists say flora, they often just mean bacteria.

The resident microbes (often called normal *flora*) get food and shelter, stable temperature, and protection. The host gets food too (because the microbes can digest and make available nutrients and vitamins that the host can't). The host also gets protection from disease, thanks to the residents' ability to compete successfully with invading microbes.

Microbiologists know that there is a vast universe of microbes about which they know almost nothing, and many reside in or on the human body. Who these bugs are and what they are doing is mostly a mystery.

Researchers do know that the nature of these associations can change depending on location. A bug can be harmless in one part of your body and deadly in another. Life-threatening peritonitis results when the abdominal cavity is invaded by bugs that shouldn't be there, usually ones that have leaked out of the intestinal tract.

Many potential pathogens do no harm unless they move in on a host whose immune system is compromised. These opportunistic infections are a worrisome feature of AIDS. There are even some parts of your body that are germ-free (or should be): your brain, internal organs, your blood, even your urine.

A Gut Feeling About Microbes

Although some parts of your body are germ-free, it is emphatically not true of the human gut—that's the intestinal tract (or GI tract) to you. It's Microbe City, the biggest we've got, and contains up to 100 trillion microbes. That's 10 times more microbes in your gut than there are cells in your entire body. The experts say this is the highest bacterial concentration of any ecosystem on Earth. Bugs love it there, and so do archaea and microbial eukaryotes. And why not? Your gut flora get protection, plus oxygen-free housing and regular meals.

You should love them, too. They also make food from your diet available to you, supplying you with nutrients from carbs such as cellulose and also from proteins that you could never digest on your own.

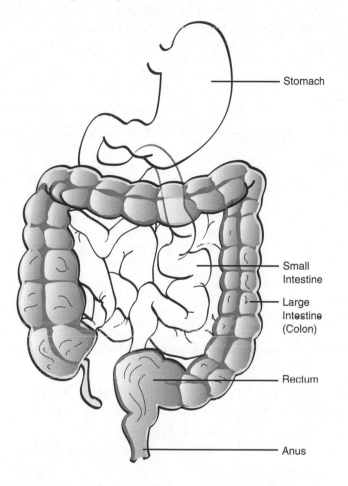

Stomach

Small Intestine

Large Intestine (Colon)

Rectum

Anus

A schematic drawing of the human intestinal tract, probably the most complex microbial ecosystem in the body. Most gut commensals live in the large intestine. The stomach is very acidic and harbors few microbes.

Researchers have recently discovered that *Bacteroides thetaiotaomicron*, a big bacterium abundant in our guts, has several metabolic pathways that can break down carbs that we can't. These enzymes are present on the bug's surface, so that it not only frees up carbs for absorption by its host, but also for neighbor microbes. *B. thetaiotaomicron* possesses more genes to deal with carbs than any other organism so far examined. That makes this bug's lifestyle extremely flexible and able to adapt quickly to hard times or other changes in its host's diet. Members of the highly specialized bacterial genus *Bacteroides* live exclusively in animal guts.

You keep yourself fed because you have the use of many other genomes besides your own. Germ-free rodents developed for research, which have no flora, need 30 percent more calories than everyday germy rats. Their gut microbes process enough of the host's food to make a significant contribution to host nutrition, an indication that the same is true for you.

> ### Little Did You Know
>
> Moreover, gut microbes make essential Vitamin K, B12, and other B vitamins that you don't make yourself.

But you get a lot more out of microbes. Normal flora out-compete pathogens for space and resources and make toxins that keep foreign invaders away from you. Normal flora also stimulate your immune system to make antibodies that attack potential pathogens, too.

The Stanford Project

Microbiologists regard human endogenous intestinal microflora as almost an organ that not only helps you nourish yourself, but also helps regulate your immune system's reactions to foreign invaders. And yet, although the animal gut is the most-studied locale in microbiology, much about it is still unknown.

Until a short time ago, scientists couldn't study microbes unless they cultured them first. And because many microbes resist culturing, they remain mysterious. However, using new tools, such as molecular fingerprinting and genome sequence analysis, a project at Stanford University has undertaken a long-term study of microbes in the human intestinal tract.

Stanford's first survey, published in 2005, concentrated on the large intestine, which is known to possess the richest diversity of microbes, mostly anaerobes.

Researchers at Stanford examined six sites in the lower digestive tract and feces of three healthy people. They expected to find new microbes, but to their astonishment they found that two out of three of the bacteria they encountered in this ecosystem were unknown to science—a completely mysterious highly diverse microbial ecosystem.

Stanford researchers also discovered that the microbes in a person's intestinal tract might be different from the microbes in that same person's feces. They sampled different parts of the intestinal tract as well as feces and found nearly 400 different kinds of bacteria, and only one archaea. Eighty percent were species that have never been cultivated in the lab.

The researchers cautioned that their samples were incomplete because no microbial ecosystem has been completely sampled yet. They also estimated that continued sampling would yield at least 500 species, probably more.

The researchers found that their three human subjects—all of them healthy—displayed somewhat different populations of intestinal microbes. They point out that a subject's genes and diet probably have a lot to do with internal microbial communities, but the specifics of how those factors shape microbial ecosystems are still largely unknown.

They also discovered that each subject had different microbial communities in his intestine and feces. This suggests that the microbe population in any one person may be as unique as a fingerprint. So unique, the researchers speculated, that a person's microbial ecosystem may even have forensic applications—for example, providing evidence of travel to particular places or the use of antibiotics.

This was a big project that took several researchers a year to complete, but it still does not provide anything like a complete picture of the intestinal ecosystem in any one person—let alone the billions of people on the planet. The researchers point out that their study is the proverbial tip of the iceberg.

One implication is that much of the information in this book is only the beginning. There is a staggering amount still to be learned about the microbial world—and plenty of work out there for tomorrow's microbiologists.

Meanwhile, In the Stomach ...

Most gut microbes live in the large intestine. The human stomach harbors few microbes because it is so acidic, which is a defense against potentially harmful swallowed microbes. There is one big exception, however. It is a microbe that has grown increasingly famous: the bacterium *Helicobacter pylori*.

Stress and spicy food cause stomach acid, and stomach acid eats away at the stomach lining and causes ulcers, right? Wrong! Only recently has medical science come to grips with how mistaken it used to be about why people develop ulcers. Almost all stomach ulcers are the result of infection by the bacterium *H. pylori*, and they can be cured with just a week of antibiotic treatment.

The human stomach is essentially a half-gallon bag of concentrated hydrochloric acid and digestive enzymes. *H. pylori* has developed unique tactics for surviving there. It corkscrews through a thick layer of protective mucous into the stomach lining beneath with the help of its spiral shape and several flagella. And then it manipulates local biochemistry to create a microenvironment for itself that has a near-neutral pH.

Scientists now know that the majority of people are home for this bug, but most display no symptoms. Poorer people and older people are more likely to be infected than the rich and young.

Only one out of five people infected with *H. pylori* develops ulcers. No one knows why, although it may be that gene differences make some strains of *H. pylori* more virulent than others. Gene differences between people probably matter too.

Some researchers argue that because most people are infected, the bug is not a true pathogen, but rather at worst a mostly harmless commensal—and possibly even provides some (as yet unknown) benefit. It's a confusing picture, especially because *H. pylori* is now suspected of causing stomach cancer.

If everybody got antibiotics, ulcers might vanish from the earth, and perhaps stomach cancer too. However, that would create more antibiotic-resistant *H. pylori*, perhaps losing whatever benefit the bug might bestow on the rest of us.

Wee Warnings

It's true that a few scientists have deliberately infected themselves to prove their theories, especially a century or so ago. But today this is definitely not an approved research technique. Kids, don't try this at home!

Barry Marshall, the Australian scientist who codiscovered the bug and the relationship, chose a time-honored method to demonstrate that *H. pylori* causes stomach disease: he drank a cocktail of it. Marshall promptly came down with gastritis (inflammation of the stomach wall, a precursor to ulcers), which got better after he took antibiotics. That result was a boost for the theory, but the link was strengthened when epidemiological studies showed that ulcers were common only in those infected with *H. pylori*. He won the Nobel Prize in physiology or medicine in 2005.

The Rest of the Tract

The population of microbes is also relatively sparse in the upper GI tract, the small intestine. Traffic moves so swiftly there that bugs get swept away by the flow before they can colonize, so there are a mere 10^5 or 10^6 organisms per ml. of fluid.

There's variation in kinds of gut microbes between individuals, but every person possesses a population diversity that is pretty stable. That implies that people's usual residents are good at dealing with outsiders. It also suggests that our own genes help shape the composition of each person's characteristic population.

In addition, genomes from more than 1,000 different viruses have been found in human feces. That's a hint that phages regularly attack gut bacteria, contributing significantly to regulating the microbe population as well. You got your first GI microbes from Mom at birth. But there are differences between infant microbe populations, depending on where a baby is born. For example, differences can occur in a developed or developing country or even in different hospital wards.

And then there's diet. It has long been known that feces flora in breastfed babies is quite different from that in bottle-fed babies. The newborn gut is germ-free, and colonizing begins at the first feeding. This initial colonization helps shape the microbial ecosystems that people possess as adults.

Ninety percent of microbes in breastfed babies are members of the gram-positive anaerobic *Bifidobacterium* because breast milk contains factors that encourage its growth. Bifidobacteria are good at breaking down sugars and apparently also good at keeping away foreign microbes (including pathogens). Lactobacilli are also common. Bottle-baby microbes tend to be gram-negative, with a preponderance of enterobacteria. They are better at digesting proteins. These differences explain why stools of bottle-fed babies look and, um, smell so different.

The Straight Poop About Farting

Speaking of smells, for about one out of three people, the final step in polysaccharide digestion is similar to that in the cow that we discussed in Chapter 9. It takes place in the colon, where archaeal methanogens process the products of bacterial digestion into methane. Methane production tends to run in families, but it's probably not genetic. Instead, it appears to result from colonization by methane-producing microbes that are passed around among family members.

The methane helps produce flatulence, also known as farting. Every day each of us releases up to four pints of flatus, as the gas mixture is called, much of it through burping rather than farting. Methane is only one component of microbe-produced flatus, which also includes hydrogen and carbon dioxide generated by bacterial fermentation.

> **Little Did You Know**
>
> Your colon contains an estimated 10^{13} bacteria per ml. In fact, bacteria account for as much as two pounds of your body's weight.

These digestive gases comprise only about 10 percent of flatus. Most flatus is nitrogen and oxygen that you've breathed in or swallowed. All these gases, including methane, are odorless. The characteristic stink of farts comes from other compounds created during microbial digestion, especially sulfur released from proteins.

Microbes on Skin

Underarms, the groin, and between the toes tend to be moist spots and harbor the preponderance of skin microbes, often gram-negative. Your skin is mostly dry and acidic. Dry and acidic are conditions most microbes don't like, but a few do live there.

Frequently cited as a harmless commensal is *Staphylococcus epidermis*, although it can be an opportunistic pathogen. There's not much oxygen in the pores of skin, so some anaerobes have moved in.

Propionibacterium

The most notorious inhabitant of pores is *Propionibacterium acnes (P. acnes)*, and from its name you can probably figure out why it's notorious. *P. acnes*, a slow-growing gram-positive anaerobe, is responsible for a good deal of misery. This bug can even (rarely) cause fatal infections, and it paves the way for serious pathogens, such as *Staphylococcus aureus*. But people don't die from zits—although sometimes they might wish they would.

Work on the *P. acnes* genome has shown that it can do well in many different environments—it can grow under microaerobic conditions, for example. It also contains many genes that help it survive on and colonize human skin. The bug makes skin-degrading enzymes and also triggers human immune system responses

Acne vulgaris—more commonly known as pimples—afflicts 8 out of 10 U.S. teens, but it is treatable these days. Try to look on the bright side: *P. acnes* takes up pore space that might otherwise be occupied by truly lethal pathogens.

Herpesvirus

There are few viruses among normal human commensal flora. Herpesvirus is regarded as one possibility, but it is often a pathogen as well. This virus is another microbe whose behavior on the skin can be annoying, although it usually is not dangerous. There are exceptions: in newborns, herpesvirus infections can cause blindness and mental retardation and can sometimes be fatal.

There are eight main types of human herpesviruses, which are double-stranded DNA viruses with big genomes. Most people acquire herpes infections as children. The virus's infection pattern explains why it often causes no symptoms at all.

Herpesvirus virions are broken up after entry to a host cell, freeing their DNA to move into the nucleus. When the virus enters nerve cells only a few viral genes are transcribed. Known as latent genes, they permit the virus to stay around for a long time without doing much of anything or producing any symptoms of sickness in the host.

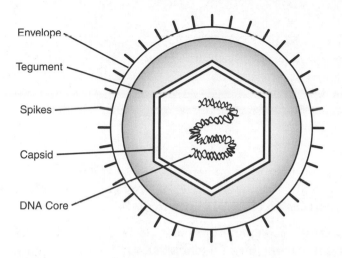

Envelope

Tegument

Spikes

Capsid

DNA Core

A schematic drawing of a typical herpesvirus. The virion core is a single large linear molecule of double-stranded DNA enclosed within an icosahedral capsid about 100 nm in diameter. It is enveloped in an outer layer made from viral glycoproteins and modified host membrane; in electron micrographs this membrane is studded with little spikes.

When reactivated in the nerves, the virus replicates and travels to the mucus membrane where it enters the body. The virus begins to replicate other genes. These genes, called lytic genes, boost the virus' rate of replication and usually kill the host cell. The genes often trigger host symptoms as well, such as fever, sore throat, or a rash. It is still a mystery why the virus wakes up and goes to work in this way, although it may correspond to the host's level of stress. Herpes simplex virus Type 1 (HSV-1) and Type 2 (HSV-2) both cause cold sores and genital herpes. HSV-1 especially likes the mucous membranes around the mouth and from there moves into nearby nerves. This latent infection often lies dormant for years in the nerves. When it suddenly becomes reactivated, it produces new virus particles and causes cold sores around the mouth.

The virus is around all the time in the majority of people, but only some are particularly susceptible to developing cold sores frequently. Because the virus is omnipresent, some scientists argue that HSV-1 and other latent viruses should be considered commensals, part of normal human flora, not pathogens.

HSV-2 prefers genitals. It causes lesions there—sometimes resembling cold sores—and also flu-like symptoms. These viruses resist the human immune system by holing up in the nerves during their latent periods. But the host can still transmit the viruses to others. Use of antiviral drugs and condoms reduces the transmission rate, but doesn't eliminate it. There is no cure, although there is hope for prevention via a vaccine.

Other herpesvirus diseases are covered in more detail in Chapter 14.

The Eyes

A few bacteria are present in healthy eyes, although there aren't many microbes in the eyes as a rule because they are washed continually by tears that contain lysozyme, an antibacterial compound.

Big Mouth

The human mouth may turn out to be a microbial ecosystem that is even more complex than the human gut. More than 700 species of mouth organisms are known, and at least half have not yet been successfully grown in the lab.

The mouth is a particularly good example of the importance of community in microbial life. In the mouth, the community is a biofilm known as dental plaque. Dental plaque gets bad press, but it has some virtues. For example, it offers some protection against pathogens from outside.

Like other microbe communities, plaque is dynamic, with different species present in varying numbers depending on local conditions. Microbial populations in plaque are about as dense as in the gut, often with more than 10^{10} bacteria per milliliter of fluid. As in any biofilm, this means cooperation, competition, and horizontal gene exchange.

The first colonizers to show up on a clean tooth are members of *Neisseria* and streptococci. These early settlers change aspects of the ecosystem, such as pH. Other organisms find conditions to their liking and move in. When plaque is thick enough to protect them from oxygen, anaerobes take up residence.

After plaque forms, it remains fairly stable in composition because competition between the microbes reaches *homeostasis*. Dental disease is believed to be the result of an imbalance between microbes. As is true of other microbial ecosystems, most mouth organisms are harmless to the host. Some, of course, are not, which helps make the human mouth a major public health problem.

def•i•ni•tion

Homeostasis, from Greek words meaning "similar" and "status" or "standing," means adjusting to changes in the external environment in order to maintain a stable internal environment.

Experts say that tooth decay is probably the world's most common disease. That represents a lot of pain and, in countries that can afford treatment, a lot of money. In the United States, dental infections are estimated to be the third costliest medical expense, ranking behind only heart disease and cancer. The government estimates that U.S. residents spend $60 billion every year treating tooth decay and $5 billion a year on gum disease treatment.

Teeth are all but eternal. Human fossils millions of years old still have intact teeth. Yet a few species of busy little mouth microbes can do serious damage to teeth in just a few years. Nearly 8 out of 10 U.S. kids have at least one cavity by age 17, and as recently as the Korean War, tooth decay was the most common reason young men were rejected for military service. Things have improved a lot, thanks largely to dental hygiene and fluoride in water supplies and toothpaste. In the 1950s, most people in the United States over the age of 65 had no teeth left at all. Only 3 out of 10 seniors are entirely toothless today. A recent government survey showed that the majority of older people still have most of their teeth, although one out of four have advanced gum disease.

Caring About Caries

Most tooth decay (also known as caries and, informally, cavities) begins with infection by a bug called *Streptococcus mutans*, which loves sugar. In fact human teeth were in pretty good shape until people fell in love with sugar just a few centuries ago. After *S. mutans* turns your dietary sugar into lactic acid that dissolves tooth enamel, other organisms, lactobacilli in particular, move in to the excavation and continue digging away.

Little Did You Know

Like all microbial topics, much is unknown about the microbiology of the mouth. Dental researchers have learned a lot from studying accumulated plaque in the mouths of students who held off brushing their teeth for three weeks. Wonder what their social life was like?

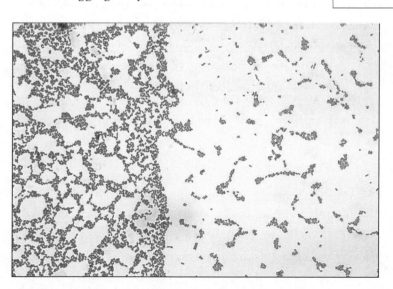

A micrograph of Gram-stained Streptococcus mutans, a major cause of tooth decay.

(Courtesy of the CDC, Dr. Richard Facklam.)

The bugs in plaque eat saliva, dead cells, and dead microbes, but what they mostly eat is what you eat. Especially sugar, which they can seize on quickly and ferment before it is all down the hatch. That's especially the case when it is consumed many times a day, because that keeps plaque in a perpetually acid state. Sugar fermenters can then increase much more quickly than the nonfermenters.

Caries have grown less frequent thanks to fluoridated water and toothpaste, and fluoride treatments in the dentist's office. Among other protective activities, fluoride kills plaque organisms. And just as your mom always said, cutting back on sugar, especially between meals, helps too. Don't feed the animals!

Gum Disease: Plenty of Periodontitis

The early stage of gum disease is called gingivitis. It's the gum inflammation that happens when plaque accumulates on teeth. Eventually this accumulation can cause destruction of ligaments and bone, and deepening pockets around the teeth that provide a cozy home for even more organisms. Dentists call this periodontitis, and eventually it leads to tooth loss.

> **Little Did You Know**
>
> Even more surprising, researchers are finding lots of archaea in the pockets around teeth. These prokaryotes wax and wane, depending on the severity of gum disease. It's not yet established that they cause gum disease, although it seems possible that microbiologists will one day discover that archaea can be pathogens after all.

It's been tough to link specific microbes to aspects of periodontal disease, although scientists do know that microbial communities differ in healthy and diseased mouths. Anaerobes are particularly common in diseased gums. A major suspect is the gram-negative *Porphyromonas gingivalis*.

However, a recent study has shown that other gram-positive species, some of them previously unknown, outnumber *P. gingivalis* in people with periodontitis. A total of 274 different species were identified. The majority have never been cultured in the lab.

Bad Breath: Halitosis

Bad breath occurs from many diseases, some of which are not caused by microbes. Bad breath also may result from tooth and gum infections, or infections elsewhere in the body, such as sinusitis and consequent postnasal drip.

Nearly 100 microbes are known to live on the human tongue, and at least some appear to cause bad breath. The tongue is an ideal place for microbes to flourish; there's a constant food supply and not much disturbance. Not surprisingly, people with bad breath are home to a different population of tongue microbes than people with breath that is toothpaste fresh.

The Future of Dental Biofilm

Dentists almost never try to identify the specific microbes in a particular case of tooth or gum disease. Traditional dentistry treats symptoms. It has not yet come up with treatment strategies based on microbiology, especially ones based on prevention—despite the fact that tooth and gum diseases are an enormous burden to people and their budgets and are infectious disorders caused by pathogens.

Still, researchers predict that future dentistry will concentrate on getting rid of the harmful microbes—for example, by depositing antimicrobial compounds directly into microbe abodes in the pockets around gums.

The U.S. government has launched a massive genome project to identify all the microbes in dental biofilm, and to figure out which ones are present in diseased but not in healthy mouths. It involves some of the same Stanford researchers that are working on gut microbes. The result could be much more specific treatments, such as mouthwashes that target only certain organisms.

Bacteria may cause human dental problems, but perhaps they will help solve them as well. Biotechnologists are trying to devise genetically engineered bacteria that can prevent tooth decay. They have stripped some *S. mutans* of their power to make acid and hope that, when applied to the teeth, the new bugs will outcompete the acid-producing bugs for life.

Airing Out the Respiratory Tract

The nose is a major portal for entry of microbes into the body. Its mazelike hairy anatomy helps keep microbes out of the respiratory tract. But the nostrils are a busy place. *S. epidermis* hangs around, and about one in five people also harbors *S. aureus* there. As you'll recall from Chapter 12, this bug is now an important pathogen because it is a reservoir of antibiotic resistance.

There are harmless microbes in the upper respiratory tract (and sometimes harmful ones, too, of course). But in health the lower respiratory tract is largely microbe-free.

The lungs are protected in a number of ways. For example, cilia in the lungs brush away mucous along with the microbes that get stuck in it. Microbes are also kept at bay in the lungs by immune system mechanisms, such as specialized cells that swallow up invaders, plus antibodies that attack them.

Urogenital Tract

Secretions from the prostate gland usually keep the male genital tract sterile, with the help of constant washing from urine. The female urinary tract is usually microbe-free, but a constantly changing microbe population normally inhabits the vaginal tract.

The microbe population of the vagina in women of reproductive age changes from day to day—and even from hour to hour—and is strongly influenced by the menstrual cycle and other hormonal changes. These alter pH and therefore microbe populations; vaginal pH is about 7.0 before puberty and after menopause, and lower in between.

Surprisingly little is known about normal vaginal flora compared to the gut or mouth. It is assumed that these microbes maintain the population in balance and might even offer some protection against other organisms, because women taking antibiotics often experience yeast infections. This happens because the antibiotics kill off the *Lactobacillus* that is a common resident of the vagina.

The Least You Need to Know

- The human body is home to trillions of microbes; these commensals are called normal flora. The brain, the blood, and most organs are (or should be) microbe-free.

- Human commensals are benign and even useful, but a number can become opportunistic pathogens in certain circumstances, for example if the host is weakened. Most commensals live in the gut and many help digest food. Commensals also compete with invading pathogens to protect the host.

- The stomach is too acidic for most microbes, but a major exception is *Helicobacter pylori*, which causes ulcers.

- Most of the skin is too acidic and dry for microbes; a major exception are the herpesviruses, which can lie dormant for years and then cause disease.

- The mouth is an enormously complex microbial ecosystem; most mouth microbes are probably benign but a few cause tooth decay and gum disease.

Part 4

The Infectious Diseases

This section provides an overview of infectious diseases. We cover diseases caused by viruses, by bacteria, by protozoa, and by fungi. You'll also learn about diseases lurking in food and water that are worrying public health officials. And finally, the fascinating fact that several chronic diseases that we don't think of as infectious (like heart disease and cancer) can be triggered by microbes.

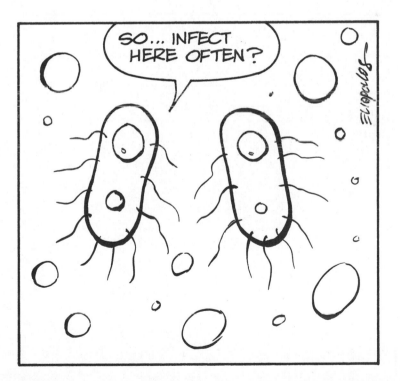

Chapter 14

Vicious Viruses: Viral Diseases

In This Chapter

- ◆ Picornaviruses cause colds and more, while hepatitis viruses cause liver disease
- ◆ Insects and other vectors carry disease
- ◆ Influenza viruses and others produce flulike illnesses
- ◆ Viruses bring about gastroenteritis and cancer
- ◆ Retroviruses trigger HIV infection and AIDS

Viruses, you'll recall, are not equipped to survive on their own. They must reside inside the cells of other organisms, hijacking that cell's machinery and turning it to their own purposes.

Viruses infect all types of life forms. In Chapter 4, we discussed one group of viruses that attack other microbes, the phage that can destroy bacteria. But viruses also prey on fungi and protozoa, to say nothing of every kind of plant and animal. By wiping out some organisms and leaving others to flourish, viruses have shaped the evolution of life on this planet, including our own. This chapter is all about the different kinds of (known) viruses.

Viruses and Disease

Most viral infections are what doctors call subclinical, meaning they don't cause obvious symptoms or disease. How do we know that? Because population studies have revealed that people possess antibodies to lots of viruses, which is evidence of past infection with them. Yet they never have had the diseases the viruses cause.

That doesn't mean these stealth infections don't matter; quite the contrary. People who are infected but not sick are a crucial reservoir of disease, since they can go about their daily lives while passing the virus on to others. Also, an infection can be subclinical but still confer immunity, so people who have the antibodies are often protected against further infection and disease. That's the principle that underlies vaccination.

Still, viruses make people sick more often than any of the other microbes. Antibiotics, despite their limitations, have done such a good job of vanquishing bacterial pathogens that viruses are now the most important causes of infectious disease in most places.

Many of these ailments are mild, of course. But a great number, especially in infants and children in less-developed countries, can be serious or even fatal and leave those who survive the infection crippled or otherwise permanently debilitated.

Little Did You Know

It is beginning to appear that diseases we used to think of as not having infectious origins, such as several disorders of the immune system (known as autoimmune diseases) and neurological disorders may actually be due to viruses. The case is not yet fully proved, but these include rheumatoid arthritis, multiple sclerosis, and at least some diabetes, among others.

Some types of cancer also result from viral infection. Cervical cancer's relationship to the human papilloma virus is now accepted by nearly all experts. Hepatitis B eventually leads to liver cancer. In this chapter, we discuss others that are either known or suspected.

To make matters worse, we have only recently begun to realize that new viral diseases appear frequently. A significant proportion of them are exceptionally virulent. We cover some of that here and talk more about emerging diseases in Chapter 17.

Many viral infections can be prevented by basic hygiene—don't forget to wash your hands!—and vaccination. But for the majority of disease-producing viral infections, little can be done except to rest and tough it out until your immune system has swung into action and beaten the invader back. When it can. Many viruses are too wily. Some of them can lie low in the body for decades, awaiting the perfect moment to begin churning out pathogenic copies.

The late-life misery of shingles is a legacy of an insignificant bout of chickenpox in childhood. HIV usually takes several years to explode into the horror of AIDS.

But the good news, if you can call it that, is that it's in the virus' interest not to kill off its host or even damage it too badly. The virus needs the host in reasonably good shape in order to make copies of itself and to provide transport so that it can infect new hosts. As a result, most viral infections are less than fatal.

Sometimes a virus can reduce its own virulence quickly. One famous example comes not from human disease, but from introduction of the deadly myxoma virus to control rabbits in Australia. Within a few years the virus became much less virulent. The result: rabbits were sick for weeks instead of days, which enabled them to move around and transmit the virus to even more rabbits before dying. In addition, the rabbits that remained tended to resist the worst effects of the virus, which now kills only about half of the rabbits it infects.

Following is a brief glimpse of many of the viruses that make us sick.

Picornaviruses: Colds and More

Picornaviruses are small RNA viruses. Many (like the ones that cause foot-and-mouth disease) affect animals almost exclusively, and the ones that infect people do not always make them obviously sick. Why, then, are we putting them first? Because these viruses are among the most common sources of human disease.

There are five main genera in the Picornaviridae family of viruses. The three picornavirus groups that infect humans include …

♦ Rhinoviruses, which cause colds, the most common human viral infection.

♦ Enteroviruses, which cause polio and other, less serious diseases.

♦ Hepatoviruses, which cause hepatitis A.

Rhinoviruses

Medical researchers say there's no good evidence that getting wet or cold causes colds, even though colds mostly occur during winter. More than half the cases of "the common cold" are due to rhinoviruses. The other half are caused mostly by coronaviruses, parainfluenza, and respiratory syncytial virus (RSV). We talk about these later, but suffice it to say that many kinds of viruses can cause colds or cold-like symptoms.

There are more than 100 different types of cold-producing rhinoviruses, so you can understand why a vaccine that can prevent colds will not happen any time soon. Rhinoviruses are transmitted through the air or by contact and usually do their infecting through mucous membranes in the nose, sometimes the eyes. They don't care for normal body temperature, which is why they dwell in cooler spots like the nose.

As you know from experience, we all get colds, some of us get several every year (especially when we are little kids), and most of them are mild. The accompanying sneezes and runny nose are produced not by the virus itself but by local inflammation, an immune system response to the virus. Infections that begin as colds sometimes go on to become something more: respiratory and ear infections, sinusitis, bronchitis, asthma, and (mostly in infants) pneumonia. Rhinoviruses don't usually cause sore throats, but some other cold viruses do.

Enteroviruses

The enteroviruses are said to be the second most common source of human infection after the rhinoviruses, but don't always generate symptoms. They appear to be the cause of a great many diseases, including episodes of "just not feeling well" (the medical term is "malaise") that we all experience from time to time and usually try to ignore. Enteroviruses are behind many fevers and undiagnosed infections of respiratory and GI tracts, eyes, mucous membranes, and skin. They also cause more serious diseases of the brain, muscles, heart, and liver.

The best-known enterovirus is poliovirus. However, since an intensive 1990s vaccination campaign, the disease (also called poliomyelitis or infantile paralysis) has all but disappeared. It is hard now to convey the terror polio inspired half a century ago, when hundreds of thousands of people got this disease every year, and many of them became paralyzed. As recently as 1988, the World Health Organization (WHO) reported 350,000 cases. Now a thousand cases per year for the entire world is the norm.

Most of the dozens of nonpolio enteroviruses that cause human disease are Coxsackie viruses, but few are serious. Some Coxsackie viruses can cause an autoimmune disease that destroys the pancreas cells producing insulin, one of the causes of diabetes.

One odd example of the work of Coxsackie viruses is hand, foot, and mouth disease, a common ailment of human infants and children. Don't confuse it with foot and mouth disease, an ailment of livestock caused by a different and unrelated picornavirus. Symptoms of the human disease are fever, mouth rash, and blisters on the bottoms of hands and feet. But infection can occur with no symptoms. Occasionally, the infection may end in meningitis or even fatal encephalitis.

Hepatovirus and Other Hepatitis Viruses

As we said, one kind of picornavirus is the hepatoviruses that cause hepatitis A. But there are so many causes of hepatitis that it makes sense to treat them all in one place.

Hepatitis, which simply means "inflamed liver," is in fact an umbrella term for a number of diseases with this warning sign but different causes. Symptoms also include jaundice (yellowing of the skin and eyes), dark urine, GI troubles (nausea and vomiting, diarrhea), fever, and exhaustion, and they can be either acute or chronic. If acute hepatitis segues into chronic hepatitis, these symptoms may disappear.

Many cases of hepatitis are due to viral infections, but it also can be the result of …

- ◆ Toxins—ethanol poisoning from heavy drinking; a great many drugs, both medical and recreational; and the death-cap mushroom.

- ◆ Metabolic disorders such as Wilson's disease and metabolic syndrome.

- ◆ Bile duct obstruction by cancer or gallstones.

- ◆ Autoimmune disease.

Five viruses from different virus families have been identified as causes of hepatitis. Some kinds of hepatitis are far more serious than others. They are designated alphabetically: hepatitis A, hepatitis B, and so on.

Hepatitis A

Some kinds of hepatitis are huge public health problems. Hepatitis A is not one of them. The cause of hepatitis A, a common but not very serious ailment, is hepatoviruses. Hepatitis A was the first hepatitis formally recognized, and originally was called infectious hepatitis. It is only rarely fatal. There is no treatment except rest and time. And here's a silver lining: hepatovirus infection generates antibodies that protect the host for life.

Vaccines against hepatitis A have been available since the mid-1990s, and the incidence of hepatitis A is falling in some parts of the world. As we write, routine vaccination of children in the United States and Israel appears to be reducing this disease in all age groups in the two countries, thanks to the herd immunity principle.

Like enteroviruses, hepatitis A is transmitted by what the medical folks delicately call "the oral-fecal route." That means the microbes are transported into people's mouths on feces, which they presumably consume unintentionally, often in water contaminated with raw sewage. The microbes pass into the gut, are shed into the new host's feces, and so are delivered into the environment where they can infect new hosts.

Tiny Tips

Many cases of hepatitis A have been passed along to unwitting diners by infected restaurant workers who failed to wash their hands after visiting the bathroom. Most cases of hepatitis A (in fact, most cases of all diseases transmitted by the fecal-oral route) can be prevented by simple hygiene. This includes hand washing with soap and warm water before preparing food or eating (and after sex!); clean, disinfected bathrooms; and clean drinking water.

Hepatitis B

By contrast, hepatitis B is a major public health concern around the world, especially Southeast Asia, where it causes an estimated one million deaths yearly from cirrhosis and cancer. This disease used to be called serum hepatitis, the form of hepatitis often passed along by users of injected recreational drugs who share needles. Hepatitis B is the major reason for developing needle-exchange programs.

Electron micrograph of hepatitis B virions. Hundreds of millions of people are infected by hepatitis B.

(Courtesy of the CDC.)

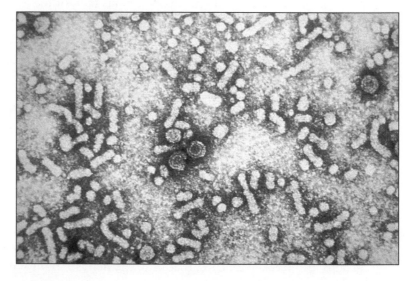

Transmission also has been traced to blood transfusions (although blood supplies are now screened for this virus), tattooing and other blood contact, sexual contact, and from mother to baby during birth. Immune system reactions contribute to symptoms.

The hepatitis B virus is a member of the family called the Hepadnaviridae, which are double-stranded DNA viruses; it's the only one that causes human disease.

Hundreds of millions of those infected with hepatitis B go on to chronic infection and liver disease and may develop immune system disorders. Antivirals can control infection in about half the cases (but not cure it). There are different genetic subtypes of the virus with different geographic distribution, but fortunately all the subtypes possess one epitope in common that is an excellent target for a vaccine; as a result, there is an effective vaccine.

Hepatitis C

Hepatitis C formerly was called non-A, non-B hepatitis. It is transmitted through blood and so is common in users of injected drugs. It also can be transmitted from mother to newborn. There are many genetic subtypes of the virus, with different geographic distributions, but the total number of infections is estimated at 150 million or more worldwide. It causes up to 20,000 deaths annually in the United States.

Most of those infected have no symptoms. But hepatitis C viruses (HCV), which are small single-stranded RNA viruses in the Flaviviridae family, cause a serious and sneaky chronic infection that can lurk unnoticed in the body for decades before causing cirrhosis and death. People infected with HIV often are infected with HCV as well, and may die of it first. The virus is now a prime reason for liver transplants. There is no vaccine, owing in part to this virus's genetic diversity.

Hepatitis D and E

Hepatitis D is genetically stunted so it needs the help of the hepatitis B virus in order to be transmitted. As a result, vaccination against B also protects against D. It is most common in developing parts of the world where B also is common. This is a single-stranded RNA deltavirus that has not yet been assigned to a virus family.

Hepatitis E is another single-stranded RNA virus that has caused epidemics in many developing countries by contaminating water supplies. It appears to target young adults in particular and can be fatal to pregnant women. It is not much of a problem in the United States. It is a member of the genus *Hepevirus* in the family Hepeviridae.

Insect-Borne Viruses and Kin: Togaviruses and Flaviviruses

Most, but not all, of these viruses cause vector-borne diseases. They are transported by insects and transmitted to people via arthropod insect bites; viruses that operate this way are called arboviruses. Most infect birds or mammals, including livestock, which makes them important economically.

> ### Little Did You Know
>
> Arboviruses are arthropod-borne. Arthropods are invertebrates with jointed legs, a group that includes spiders and lobsters as well as mosquitoes and other insects.

In people, these arboviruses cause various kinds of encephalitis (inflammation of the brain) and diseases involving fever, rashes, and often severe joint pain. The best known in this country is probably West Nile virus, a media darling that has wiped out bird populations but is not, numerically speaking, an important human disease. The most common is dengue. More on these in a moment.

Arbovirus diseases can be life-threatening illnesses and have long-term consequences. They are a public-health problem in the world's warm places. A few are so contagious and potentially dangerous that special safety measures are required for handling them in labs.

These RNA viruses are also small, but about twice the size of picornaviruses, about 60 nm. They are wrapped in envelopes with at least one glycoprotein type on the surface that helps them to be recognized by and adhere to cells. Each viral type tends to be restricted in geographic reach.

Alphaviruses (members of the virus family Togaviridae) are transmitted by mosquitoes. More than two dozen disease types have been identified. Infection does not always lead to serious disease. Even encephalitis can be mild, but it also can mean seizures, coma, and death. The most common form in the United States, St. Louis encephalitis, is in fact not very common at all, well under 200 cases per year. Like many of the alphaviruses, it mostly preys on birds.

Flaviviruses (members of the virus family Flaviviridae), also transmitted by mosquitoes and sometimes ticks, cause many kinds of encephalitis, plus dengue and hemorrhagic fevers, and possibly hepatitis C. Some 70 flaviviruses have been identified.

Dengue

Dengue fever is the most common viral disease transmitted by mosquitoes, with perhaps 100 million cases yearly. Symptoms include high fever, severe body aches, nausea and vomiting, and rash. It appears to be a newer disease; the first epidemics were reported late in the eighteenth century. The CDC says its worldwide distribution is similar to malaria, and is now a major public health problem in the less-developed countries.

Dengue hemorrhagic fever, the more serious form, first appeared as an epidemic in the 1950s in Asia. Hundreds of thousands of cases are reported every year; the death rate is about 5 percent, mostly among the young. As the name suggests, in this disease small blood vessels grow leaky, causing easy bruising, bleeding from gums and nose, and perhaps internal bleeding. The result can be circulatory collapse and death from shock.

Dengue has one particularly horrible trait. Surviving a bout of dengue not only does not protect against another, it appears that it can even make a second episode worse, a characteristic called immune enhancement. The disease is caused by four related viruses, each of which produces distinct host antibodies. When any one kind of antibody binds to the virus, it actually can add to the virus's ability to enter a host macrophage, where it makes many copies of itself. Many cases of dengue hemorrhagic fever are thought to be due to this mechanism.

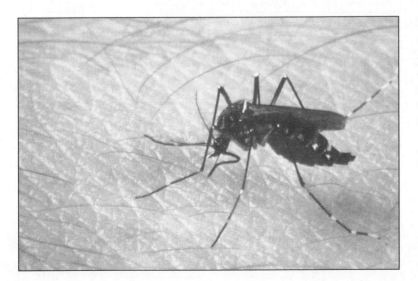

This Aedes aegypti *mosquito, primary vector for dengue, rests on the skin of a volunteer during her blood meal. If she is infected with the dengue virus, she can transmit it to a new host.*

(Courtesy of the CDC, Robert S. Craig.)

People can protect themselves with mosquito nets and repellents, which is wise because there is no specific dengue treatment. Nor is there a dengue vaccine, and none is expected soon. Control measures, when they exist at all, consist largely of destroying mosquito-breeding places, a long-term strategy. Worst of all, new dengue virus strains are expected to emerge.

Yellow Fever

You'll recall from Chapter 4 that yellow fever was the first disease known to be caused by a virus, and that it's not much of a health problem in the United States. There is a good vaccine, but yellow fever is still a serious problem in South America and Africa. The virus causes an estimated 200,000 cases, and 30,000 deaths, annually.

The disease is a devastating hemorrhagic fever, hence one of its other names: black (that is, bloody) vomit. Yellow fever is important historically. It wiped out an army of 40,000 that Napoleon sent to quell rebellion in Haiti and forestalled attempts to build the Panama Canal until researchers invented a vaccine that could protect workers.

West Nile Virus

Four out of five people infected by West Nile virus *have no symptoms at all.* Most of the rest suffer some aches, fever, nausea, and perhaps a rash for a few days. Severe illness—high fever, encephalitis, meningitis, convulsions, coma, paralysis—develops in only one in every 150 infected people, most of them elderly or ailing. Death is uncommon.

In short, the human disease that West Nile virus causes isn't all that serious compared with many other infections. Why, then, are people so worried about it? Because it appears to be an exceptionally fast-moving, emerging disease. First reported in Africa in 1937, it arrived in the United States (in New York to be exact) in 1999, yet by 2005 was found all over the country. It is now responsible cumulatively for more than 16,000 cases of encephalitis in the United States, and uncountable deaths in other species, especially birds.

The future of West Nile virus in the United States is anything but clear from its very short history. Some experts think it may establish a pattern similar to that of the St. Louis encephalitis virus, its close relative. That pattern is extremely variable, with small local

Wee Warnings

West Nile virus almost always is transmitted to people by mosquitoes that have fed on infected birds. It cannot be passed along to another person by everyday direct contact, although (very infrequently) it has been spread through blood transfusion, organ transplant, breastfeeding, and pregnancy.

epidemics one year and only a few cases the next. West Nile may be more common, however, because it infects many more mosquito species and is more likely to cause symptoms. Risk of disease in individuals, however, is likely to remain low.

Rubella

We discussed rubella epidemiology and vaccination in Chapter 11. Rubella is a togavirus, but, unlike most of them, is not transmitted by an insect. It's spread by close contact.

Bunyaviruses: More Arboviruses Plus Hantavirus

With one known exception, hantavirus, the 50-plus human disease-causing members of this group of single-stranded RNA viruses are also arboviruses. They belong to the Bunyaviridae family and cause fevers and hemorrhagic fevers, aches, nausea, encephalitis, and sometimes death.

La Crosse encephalitis, transmitted by a woodland mosquito, is the most serious bunyavirus disease in the United States. It is uncommon and rarely fatal. Rift Valley fever, also carried by mosquitoes, causes massive epidemics in African livestock. Human epidemics appeared in Mauritania in 1987 and Egypt in 1977.

Hantavirus is not transmitted by insects but rather by rodents, mostly rats and mice. The species identified originally in Korea causes a disease known as Korean hemorrhagic fever, but there are other hantaviruses that cause different diseases.

In the United States, the best-known hantavirus is the Sin Nombre virus, cause of hantavirus pulmonary syndrome. It results in severe respiratory symptoms that lead to death in about half the cases, but is not very common. This virus was first isolated in the Southwest. Humans usually become infected by breathing air that contains the virus, or by touching or eating material containing the virus, which rodents carry and shed in their feces, urine, and saliva.

Arenaviruses and Rodents

Arenaviruses, in the Arenaviridae family, another group of single-stranded RNA viruses, also cause diseases transmitted by contact with rodent excreta. They don't give the rodents trouble and mostly cause only flulike symptoms in people. But infections sometimes move on to neurologic disorders or hemorrhagic fevers. The best known is probably African Lassa fever.

Another Vector-Borne Disease: Rabies

Rabies is caused by a Rhabdovirus, a group of more than a hundred viruses that infect all sorts of animals and plants. They are single-stranded RNA viruses in the family Rhabdoviridae. Rabies is the most important human disease among these, but another member of the family, vesicular stomatitis virus (VSV), is a popular model for studying viruses and their evolution. VSV is an arbovirus, transmitted to mammals via insects, but is thought to be harmless to people.

Rabies viruses are anything but harmless. They infect mostly carnivores and cause virulent encephalitis that nearly always is lethal in unvaccinated humans. The virus is in saliva, usually conveyed to a new host via a bite from an infected one. Infection makes a host exceptionally aggressive and given to biting, which helps ensure that the virus will be passed along.

Rabies has not been much of a problem in Western countries for the past 50 years, thanks to campaigns to vaccinate dogs. Today the disease lurks mostly in wild animals, including bats. Vaccine prevents the disease even after bites from rabid animals if it is swiftly given.

Influenza Viruses, Old and New

Flu is caused by single-stranded RNA viruses of the genus *Orthomyxovirus* in the family—we bet you could have guessed this—Orthomyxoviridae. These viruses strike the upper respiratory tract mostly: nose, throat, and bronchi, but not usually the lungs. A typical bout of flu means high fever, runny nose, severe achiness, and feeling extremely lousy for a week or so. Yes, the good old malaise. Flu is generally only a serious disease in the very young, the very old, and people who are already ill, sometimes leading to pneumonia and death.

There's a seasonal flu epidemic every year that hits from 5 to 15 percent of the population. The WHO estimates there are between three and five million serious cases annually and puts deaths at between a quarter- and a half-million. WHO says the burden is greater in the less-developed countries but declines to estimate numbers there.

But, as we pointed out in Chapter 11, medicine is enormously interested in this disease because it expects a new pandemic—a global epidemic—before long, and fears that it may be worse than the notorious 1918 flu. This will be a different, far more virulent disease than the annual flu. Some experts have estimated that one in three people will fall ill and hundreds of millions might die.

A vaccine will protect against this new disease, but only eventually. Once the pandemic virus appears, it will still take months to develop a vaccine against it and more months to get the vaccine into mass production.

Antivirals exist that probably would work after infection, but they must be taken soon after symptoms appear, and medical journals are already reporting that microbes have developed resistance to them. Also, the drugs are in short supply, even in the developed world that can afford them. Unfortunately, most people cannot.

The expectation is that a bird flu that has already mutated so that it can (occasionally) infect humans will then acquire genes from a human flu virus, changing into a form that can pass easily from person to person. The flu virus is made up of eight RNA segments that code for at least ten proteins. When two flu viruses, even from different host species, infect the same cell, they can effortlessly swap segments that enable each one to infect new hosts.

There have been less-serious pandemics since 1918, one in 1957 and the other in 1968. There also have been a half dozen pandemic false alarms in the last few decades. A false alarm is an outbreak in which a novel strain has jumped the species barrier but has been confined to a few people and often has not been lethal.

Of the types of influenza viruses, only two trouble *Homo sapiens*. They are known as A and B; A causes more serious disease and mutates more often. Subtypes of A are defined by variations in hook-like antigens, hemagglutinin (H) and neuraminidase (N), on their surfaces. The two currently important human subtypes are H1N1 and H3N2. Mutations in the genes that code for these antigens permit the virus to evade the host immune system and allow it to be passed easily from person to person.

Most regional epidemics are due to type B. But human pandemics, such as the 1918 flu are caused by type A. The 1918 virus, believed to have transferred originally from birds, is said to have killed twice as many people as died in World War I and was particularly deadly to young adults, a very different pattern from annual flu. The Asian bird flu, subtype A H5N1, is being monitored by medical authorities for signs of mutations that will speed its passage between humans. In the (comparatively) few humans it has infected so far, the death rate has been 50 percent.

More Viruses That Cause Colds and Flu-like Illnesses

As we said earlier, a great many different viruses cause respiratory diseases, including colds. Here are some more of them.

Paramyxoviruses

Several groups of viruses in the Paramyxoviridae family cause human disease. As you might expect from the name, parainfluenza viruses, another single-stranded RNA virus, cause respiratory infections. They are believed to be responsible for a third or more of such infections in infants and young children. Mostly they resemble mild colds with fever, but occasionally go on to become croup, bronchitis, or pneumonia, possibly fatal. They are passed along mainly via small children, by direct contact or through virus-laden droplets in the air. There is no vaccine yet.

Tiny Tips

As you know from Chapter 11, vaccines work well against mumps and measles. These two paramyxovirus diseases are now uncommon in developed countries.

Respiratory syncytial virus, another paramyxovirus, is thought to infect just about everyone before the age of two. It attacks primarily the upper and lower respiratory tract and the middle ear, causing otitis media. Mostly it appears to be just another cold, but it can be fatal in the very young and very old. Because it is essentially universal and reinfection is common, RSV is burdensome to the health care system and the economy. There is no vaccine yet for this one, either.

Coronaviruses and SARS

Another group of single-stranded RNA cold viruses, coronaviruses (family Coronaviridae) are thought to be responsible for perhaps one in five mild colds. Coronaviruses get their name from the halo ("corona") of glycoproteins that appears around them in electron micrographs. They infect birds and mammals. Until 2002, coronaviruses were of little interest to researchers because they didn't seem to cause serious disease.

And then came SARS.

Severe acute respiratory syndrome, a flu-like disease with high fever that often ends in pneumonia, emerged in China that autumn. By February 2003, it had moved to Hong Kong and later that month to North Vietnam. By March it was in Bangkok and other Asian cities and by April in Toronto. And then it vanished just as quickly. By June the whirlwind of an outbreak was over. Toll: more than 8,000 cases in 27 countries, some 800 dead. In the United States, eight cases, no deaths.

SARS turned out to be caused by a new coronavirus, one that, since the outbreak ended in 2003, apparently no longer exists except in a lab. The virus may have jumped to humans from a civet, a catlike carnivore that is eaten in China, or from bats. The virus was identified within a month of the beginning of the outbreak, a record and a much-praised piece of work.

The SARS experience is expected to shape preparedness for all kinds of emerging diseases especially with respect to facilitating communications among health-care workers and the public and also to reviving standard public-health measures such as quarantine that had been regarded as old-fashioned.

Adenovirus: A Research Tool Causes Many Diseases

This big Adenoviridae family of double-stranded DNA viruses infects animals and birds; about 50 that cause human disease have been identified. Most of this is respiratory disease (colds, pneumonia, croup, bronchitis), but adenoviruses also can cause gastroenteritis and infections of the eye and urinary tract. They are passed on easily by just about any route: breathing in droplets containing the virus, direct contact, the fecal-oral route, contaminated water, and even sexual transmission.

Adenoviruses also are important as tools for medical research. We mentioned adenoviruses in Chapter 6 because they are easily modified genetically and often used as vectors for gene therapy. For the same reason they are regarded as potentially good vectors for experimental vaccines, able to convey antigens into hosts to produce an immune response.

Parvoviruses Need Help

Parvoviruses are of interest to microbiologists mostly because they are so tiny—around 20 nm—that they lack the right genes for replicating in host cells. These single-stranded DNA viruses, family name Parvoviridae, need help to make copies of themselves.

One group of parvoviruses replicates with the assistance of helper adenoviruses (often called adeno-associated viruses). Others don't need a helper virus but replicate only in cells that are already replicating their own DNA before they divide.

It is only this latter group that appears to cause disease, all of them in mammals. These are severe and, in animals, often fatal diseases. The best known is probably canine parvovirus, with symptoms very like feline distemper (from which it is thought to be descended). Canine parvo appeared seemingly from nowhere in the mid-1970s and took only two years to move around the world. It is extremely contagious but can be prevented by yearly vaccination.

The best-known human parvovirus diseases result from infection with the parvovirus known as B19. Most adults possess antibodies against B19, evidence of past infection. The most common B19 disease, erythema infectiosum, is a childhood flu-like illness with rash and temporary arthritis also called fifth disease or, descriptively, red cheeks or slapped cheeks disease. In patients already suffering from anemias such as sickle-cell anemia, B19 infection causes a severe acute anemia. Late in pregnancy B19 precipitates miscarriage or the form of stillbirth known as hydrops fetalis.

"Stomach Flu": Not Flu at All

What is the most common death-dealing infectious disease of all in infants and young children? Diarrhea.

It's known by several names, among them winter diarrhea, acute viral gastroenteritis, and stomach flu. But it's not flu at all. Real flu is influenza, the upper respiratory disease we discussed above. Flu can be prevented by flu shots, but there is no vaccine for "stomach flu."

Rotavirus: the Most Important Virus You Never Heard Of

The pathogen responsible for most of this gastrointestinal misery is a double-stranded RNA virus that, electron microscopy reveals, looks like a wheel. That's how rotavirus got its name; *rota* means "wheel" in Latin.

An electron micrograph of rotavirus virions along with other, unidentified virions. Rotavirus has a distinctive wheel-like shape, which is how it got its name. Rota is the Latin word for "wheel."

(Courtesy of the CDC.)

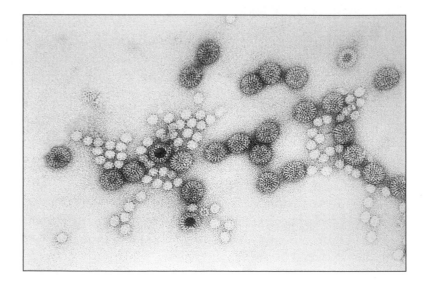

Rotavirus is a genus in the Reoviridae family. At least four groups of rotavirus infect animals. Human rotavirus comes in three types: A, B, and C. It infects almost everybody. Most people have antibodies against this virus, evidence of past infection, by age three.

In children under five, rotavirus A infection is behind at least 25 million visits to clinics, two million hospitalizations (a third of the total), and more than 600,000 deaths every year. It causes an estimated three million cases of gastroenteritis annually in the United States.

Group B rotavirus, often called adult diarrhea rotavirus, troubles people of all ages, and group B in drinking water has caused epidemics in China. Group C is rare but has popped up all over the world.

The pattern is familiar to most of us: vomiting and watery diarrhea for up to a week or so, usually with fever and bellyaches as well. In the developed world, that unpleasantness is usually followed by complete recovery and partial antibody protection against the next rotavirus that comes along.

Children in poorer countries are not so lucky. Most deaths are due to dehydration and electrolyte imbalance, so the chief treatment is oral rehydration therapy that replaces fluids, salt, sugar, and potassium.

Usual transmission of this very contagious virus is by the fecal-oral route, mostly via close contact. Contaminated water is sometimes responsible and perhaps food handlers, too.

Finding a vaccine against rotavirus has a checkered history, but in 2006, two new vaccines were shown to be both safe and effective against it. There is still a long way to go. Organizing, financing, and monitoring a worldwide campaign to vaccinate hundreds of millions of children is a project of staggering size and complexity. At this writing, however, it appears that public health workers now at least possess the basic tools to prevent this calamitous infection.

> **Little Did You Know**
>
> One reason rotavirus disease is so common is that a little goes a long way. Disease can be triggered by exposure to fewer than 100 virions. Rotaviruses also are able to elude sanitary measures that work fine against bacteria and protozoa.

Norwalk Virus, a.k.a. *Norovirus*

The other major causes of "stomach flu" aren't flu viruses either. They are noroviruses, formerly known as Norwalk virus after a 1968 epidemic in Norwalk, Ohio. They are a major cause of "food poisoning," a topic for Chapter 17, and the chief cause of epidemics of viral gastroenteritis. They are known to the general public chiefly because of well-publicized infections of cruise ships.

Retroviruses, HIV, and Deadly AIDS

What makes retroviruses retro? Retroviruses, which are RNA viruses in the family Retroviridae, do transcription backward, or at least backward in terms of the way transcription usually happens.

As we pointed out in Chapter 6, in classical genetics, DNA is transcribed into RNA as part of the cellular process of making proteins. Retroviruses reverse that step, transcribing RNA into DNA. This they do with the help of a special enzyme, reverse transcriptase. Then they insert this new DNA into the host genome, using the enzyme integrase, and become a permanent resident.

Reverse transcription has an exceptionally important genetic result: because reverse transcriptase does not include the usual error-checking procedures of DNA-into-RNA transcription, retroviruses mutate unusually often. They can develop resistance to antivirals quite quickly and also present a moving target for frustrated vaccine developers.

Endogenous Retroviruses

As a result of reverse transcription, there are two main kinds of retroviruses:

◆ Endogenous retroviruses that have been integrated into the host reproductive cells and get passed on to descendants via vertical transmission.

◆ Exogenous retroviruses, which like other viruses are passed from host to host via horizontal transmission.

Endogenous retroviruses (sometimes called proviruses) probably evolved first, from transposable elements, the jumping genes we discussed in Chapter 6. They are thought to have repackaged themselves as virions, making it possible to infect other cells and become exogenous retroviruses.

Endogenous retroviruses are a permanent part of the genomes of most vertebrates and some invertebrates, comprising an estimated 8 percent or so of the human genome. They are often part of the junk, or noncoding, DNA in a host. But some of them have become actual genes that do cellular work and are a true part of the host genome. They control gene transcription, contribute to placenta development, and even help the host fight off exogenous retroviruses.

However, endogenous retroviruses also figure in disease, including human disease. They appear to play roles in leukemias, which are cancers of the bone marrow and blood resulting in wild growth in the number of white blood cells (leukocytes). They seem also to be involved in disorders of the immune system such as multiple sclerosis.

Exogenous Retroviruses, Especially HIV

Most exogenous retroviruses are pathogenic, spreading infection by moving from host to host just like other infectious viruses. These retroviruses also cause many leukemias and sarcomas (cancers of the blood vessels, bone, fat, cartilage, and muscle) in both animals and people.

But exogenous retroviruses are best known for causing an immune system disorder that has become notorious in the past couple of decades: human immunodeficiency virus (HIV) infection, which eventually is fatal in the form of acquired immunodeficiency syndrome, AIDS. HIV infects the T cells of the immune system, which is why it is so devastating.

The HIV pandemic is one of the most destructive ever. Since it was identified in the early 1980s, HIV infection has swept the world, although it is still at its worst in sub-Saharan Africa, where 60 percent of those infected reside. The United Nations reports that some 40 million people are infected with HIV, and about five million more become infected, and more than three million die every year.

HIV infection can be asymptomatic but is more often acute (with fever, aches, GI symptoms, rash) or moved on to AIDS. AIDS symptoms include severe immune deficiency, opportunistic infections (especially pneumocystis pneumonia), cancer (especially Kaposi's sarcoma), nervous system derangements that can include dementia, and, finally, death. While the infection cannot be cured, in many parts of the world, progression to AIDS, a process that often takes several years, has been slowed even further by new drugs.

HIV is transmitted through sexual contacts involving exchange of bodily fluids, through infected blood, and from mother to child in early infancy. The clinical picture often is

Tiny Tips

There are at least two types of HIV. HIV-1 is found worldwide and is responsible for most infections. HIV-2 is present mostly in West Africa; it is believed to be somewhat less infectious and causes less serious disease.

complicated by the fact that other infectious agents (notably hepatitis C) are passed along at the same time.

There has been progress against HIV infection, but the remaining problems are horrific. Yes, there are drugs that can slow the progression to AIDS, but how can these expensive creations get to the poor people of the world who need them most? A vaccine is the highest priority but is proving to be an exasperating technical challenge. As we write, it looks to be many years away.

This electron micrograph shows both of the major disease-causing retroviruses. HTLV is the human T cell leukemia virus, and HIV, of course, is the human immunodeficiency virus that leads eventually to AIDS.

(Courtesy of the CDC.)

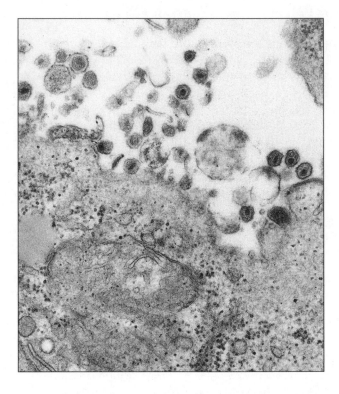

The other important group of exogenous retroviruses are the three known types of human T cell leukemia viruses (HTLV). Most of the time infection, which is lifelong, causes no symptoms at all, but HTLV also is associated (obviously) with adult T cell leukemia, which occurs in less than 5 percent of those infected. Uncontrolled growth of cell numbers commonly does not develop until three or four decades after infection, which can be transmitted by sexual contact and in blood and breast milk. HTLV also causes immunosuppression and a neurological disease called tropical spastic paraparesis. There is no vaccine yet against these viruses either.

More Cancer-Causing Viruses

We've already mentioned the carcinogenic properties of hepatitis and retroviruses. Here are some other viral causes of cancer, although they also cause different kinds of diseases as well.

Papillomaviruses

We have talked about human papillomavirus before; it's a double-stranded DNA virus with the family name Papillomaviridae. It causes the second most common cancer in women (and the third most often fatal one), cervical cancer. These viruses also cause another sexually transmitted (but not usually serious) ailment, genital warts, and many other warts and wartlike lesions as well, including plantar warts and ordinary warts. Unlike some papillomavirus infections of the genitals, mouth, and throat that go on to become cancerous, these almost always are benign growths of no medical consequence.

There are several types of papillomaviruses, and only a few, especially types 16 and 18, seem to have much cancer-causing potential. An experimental vaccine that works against these two is doing well in clinical trials as we write.

Herpesvirus

We talked about some herpesviruses in Chapters 4 and 13. Herpesviruses, which are double-stranded DNA viruses in the family Herpesviridae, share two noteworthy characteristics with some other viruses. First, herpesvirus infection is near-universal; by the age of 50, nearly everybody possesses herpes antibodies, evidence of past infection that may well have displayed no symptoms at all. Second, herpesvirus infection is incurable. Herpesviruses hang around the host in dormant states long after their initial infection, often emerging years or even decades later to cause trouble, sometimes very serious trouble.

This long-lived infection helps explain why herpesviruses can sometimes cause cancer, which usually takes a long time to develop. But it's also puzzling. Nearly everybody is infected with herpesviruses, yet only a comparative few develop herpes-related cancers.

There are more than 100 herpesviruses, but only eight are known to cause human disease:

◆ *Herpes simplex* virus types 1 and 2, which cause cold sores and genital herpes

◆ Varicella-Zoster virus, which causes chickenpox and shingles

◆ Epstein-Barr virus, a cause of infectious mononucleosis and possibly many cancers

◆ Cytomegalovirus, dangerous mostly to fetuses and the immunocompromised

◆ Human herpesvirus types 6 and 7, which cause roseola or "sixth disease," a fever and rash that are not usually serious

◆ Kaposi's sarcoma virus, which causes a distinctive cancer in the immunocompromised, especially after HIV infection

◆ B virus, which causes a very rare form of fatal encephalitis in people; mostly a rhesus monkey disease

Herpes simplex infections, like most herpes infections, are incurable and quite contagious but practically universal and often without symptoms. Host immune systems can annihilate active virus particles, but the virus flees into nerve ends where it can hide successfully from immune surveillance.

HSV-1, the more common and less serious infection, causes mostly cold sores and blisters near the mouth, whereas HSV-2 causes mostly blisters and other lesions on genitals. But both viruses can cause either disease. More rarely they cause other skin infections and also a form of encephalitis that often is fatal.

HSV is transmitted by close contact through breaks in the skin, but not necessarily sexual contact. Experimental vaccines are in clinical trials but are still some years away from wide distribution. A number of prescription drugs can control outbreaks. But, at least for now, nothing can get rid of an existing *Herpes simplex* infection.

Varicella-Zoster virus, also known as human herpesvirus-3, causes the usually benign childhood ailment chickenpox and then lurks in the host nerve fibers for life. About 20 percent of the time, HHV-3 emerges decades later to resume replicating, this time bringing on the misery of herpes zoster, commonly called shingles.

Little Did You Know

There is a vaccine to prevent chickenpox. A vaccine to prevent shingles underwent a large and successful clinical trial in 2005, although it will probably not be available for years. Commenting on the success of the trial, one expert pointed out that, given the misery of shingles, grownups may need a vaccine more than children do.

Shingles causes blisters and an itching, burning sensation and shooting pain on one side of the body. Even worse, a shingles attack is often followed by postherpetic neuralgia, severe and sometimes excruciating pain emanating from the same nerves and therefore untreatable with the usual analgesics. It can last for months.

In the United States, there are half-million shingles cases every year. Shingles is most common in those over 50, but anyone who has had chickenpox is at risk. Antivirals can reduce shingles symptoms.

The initial Epstein-Barr virus (human herpesvirus-4) infection is often symptom-free but can cause infectious mononucleosis (also called mono or glandular fever.) This is commonly a teen disease, especially in the United States, and means fever, sore throat, and swollen lymph glands.

EBV was the first oncovirus (cancer-causing virus) identified. It is now thought to be a cofactor; even in the EBV-infected, cancer won't develop unless other conditions also are present.

The strongest association is with Burkitt's lymphoma, a non-Hodgkin's lymphoma common in Africa and thought to be related to malaria's ability to reduce immune surveillance. This lymphoma often takes the form of a giant tumor on the jawbone.

Nasopharyngeal carcinoma, another EBV-associated cancer, is most common in China and is thought also to be related to genes and carcinogens in the diet. EBV is suspected of being a cofactor in many other cancers, including other lymphomas and carcinomas, plus leukemias, sarcomas, and even breast cancer.

Cytomegalovirus (human herpesvirus-5) infection, also a cause of infectious mononucleosis, is also near-universal but not usually much of a problem except in fetuses and people with weakened immune systems. CMV is the most common congenital virus infection in the United States. It can cause infant death and a variety of severe problems ranging from deafness and blindness to mental retardation. In the immunocompromised, a latent CMV infection can reactivate to cause inflammatory conditions and organ damage as well as death.

Poxviruses: A Pox Upon Them

The Poxviridae are a large family of double-stranded DNA viruses that infect many kinds of animals. We discussed the demise of smallpox, the major human poxvirus, in Chapter 4. Smallpox may be gone, but it is not forgotten, and we discuss why the experts are still worried about it in Chapter 19.

With one exception, the other poxviruses that infect humans are all really animal diseases. The exception is molluscum, which causes painless skin lesions, chiefly in children. It is transmitted by direct contact, including sexual contact, and is of little interest to medicine.

The Least You Need to Know

♦ Viruses infect all life forms causing more infectious disease than any other microbe, and while most of these diseases are mild, some can be serious or even fatal.

♦ Viral infections can be prevented by basic hygiene, especially careful hand-washing.

♦ Colds are perhaps the most common human viral disease, and they are caused by many different viruses. Some viruses cause serious disease many years after infection.

♦ A great many viral infections are transmitted to people by vectors that include other animals and insects, especially mosquitoes.

♦ There is a seasonal influenza epidemic every year, but officials are now expecting a large pandemic of a much more severe flu, caused by a new flu virus, that will be fatal for millions.

♦ Retroviruses cause some kinds of leukemia and other cancers; the retrovirus HIV now infects many millions and causes severe disease that leads eventually to AIDS and death.

Bacterial Diseases

In This Chapter

- ◆ Bacterial pathogens and their hosts
- ◆ Bacterial families: *Staphylococcus, Streptococcus, Neisseria, Bacillus*
- ◆ More bacterial families: *Clostridium, Mycoplasma, Pseudomonas, Brucella, Yersinia, Haemophilus, Bordetella, Corynebacterium, Mycobacterium, Chlamydia, Treponema, Legionella, Rickettsiae,* and others

Microbiologists are reasonably confident that, with rare exceptions, they have identified most bacterial causes of disease. Because of their enormous flexibility and adaptability, however, bacteria will always be fascinating to medical microbiology. In this chapter, we provide a brief overview of the bacteria that cause human disease.

Bacterial Pathogens and Their Hosts

Researchers have studied bacterial diseases the longest, and so bacterial diseases are the ones that scientists now know best. Some bacteria are pathogens, pure and simple. Others, the opportunistic pathogens, cause disease only when age or another disease weakens the hosts. Most bacteria never cause disease.

Vaccines, antimicrobial chemicals, and antibiotics are the weapons humanity has invented to combat bacterial diseases. But, as we have shown elsewhere (especially in Chapters 11 and 12), these weapons are imperfect—and to some extent always will be. In the race between people and pathogens, the task of keeping up is endless.

Host immune systems, of course, apply several different tactics to fight off bacterial pathogens. IgA, for example, is active on the epithelium. But the host's chief weapon appears to be the humoral immune system and its store of antibodies and complement. Here are some ways the humoral immune system works:

Wee Warnings

Decisions about whether a bug is harmful are not always final. Researchers have discovered that some bacteria thought to be harmless actually cause disease. And some bacteria that are harmless mutate and develop the ability to cause disease.

- ◆ IgG and complement coat the pathogen to attract phagocytes.

- ◆ IgG and IgM force bacteria to clump together, making them easy targets for phagocytes.

- ◆ Antibodies cooperate with complement to persuade bacterial cells to burst open.

- ◆ Complement attracts phagocytes to the infection.

- ◆ Special antibodies known as antitoxins block the toxins that bacteria produce.

That is not to say that cell-mediated immune system mechanisms are irrelevant. Here are some cell-mediated immune-system tactics:

- ◆ They are especially good at dealing with bacteria that hide inside host cells.

- ◆ They can stimulate macrophages to become even better at their jobs.

- ◆ They attract macrophages to the infection and increase the number of other immune system cells.

As you might expect, however, bacteria can be clever at combating host immune systems. For example, bacteria can modify their antigens. Many bacterial pathogen species come in several different flavors, each one with a different antigen—meaning that a single species of bug can cause the same disease in the same person many different times. That's because immune system responses must be designed from scratch with each infection, because each infection appears to be a new organism. Others, such as the gonorrhea bug discussed below, possess several different genes for a single antigen. These bacterial pathogen species can make a number of different versions of an antigen that an immune system is unable to recognize at first, even an experienced immune system that has encountered the bug before.

Some bugs know how to stimulate an immune response and then hijack the response for their own purposes. Several know how to subvert phagocytes. *Myco-bacterium tuberculosis,* for example, is notorious for using macrophages as its primary hosts and preventing its own dissolution by macrophage lysosomes. This phenomenon was discussed in Chapter 10, and other examples are described later in this chapter.

Tiny Tips

Here's another hijacking tactic. *Listeria monocytogenes* reproduces inside macrophages. It then builds bridges to nearby macrophages and crosses into them without ever emerging from the cells, which could expose them to antibodies.

The syphilis bug, *Treponema pallidum,* hides from antibodies by slathering itself with host molecules until it has crept into the central nervous system tissues where antibodies rarely penetrate.

Some bugs have figured out ways of suppressing immune responses, such as producing toxins that function as superantigens. Superantigens are proteins that latch on to antigen receptors on a sizeable number of T cells, triggering production of cytokines that shut down immune responses. Staphylococci are one example; they release a chemical that causes toxic shock, a potentially lethal syndrome. Less dramatically, some pathogenic bacteria produce enzymes that can disable or weaken antibodies.

Immune systems can also become tolerant to foreign antigens. Exposure to bacterial antigens at certain times in fetal life, for example, can lead the infant immune system to classify the antigens as self. High levels of circulating antigens can have the same effect in older hosts. Sometimes bacterial antigens are similar enough to host antigens to provoke a weak immune response or none at all. An example is some polysaccharides in bacterial capsules.

Many bugs that cause intestinal distress are transmitted to people in their food. This topic is discussed in Chapter 17.

Staphylococcus: MRSA and More

Staphylococcus, which is a gram-positive pathogen of mammals, gathers in clumps that are often compared to bunches of grapes. You have probably heard most about *S. aureus.* There are more than 30 staph species, some of which live in soil. About a dozen of them are human commensals, but only *S. aureus* and *S. epidermidis* are serious human pathogens.

Tiny Tips

Staphylococcus is one of several families of bacteria that cause suppurative diseases, which are diseases that produce pus.

Highly contagious *S. aureus* is the major cause of infection acquired in hospitals, called nosocomial infections. It is the SA in MRSA, the current epidemic of methicillin-resistant *S. aureus*. This bug is exceptionally good at becoming resistant to antibiotics. It easily acquires new plasmids and transposons and its chromosome also mutates easily. *S. aureus* has now even developed resistance to the antibiotic of last resort, vancomycin. This version of *S. aureus* is now called VRSA.

But *S. aureus* also causes many other infections of widely ranging severity and produces toxins that make people sick. Among these are the heart disease known as bacterial endocarditis; boils and other skin infections; pneumonia; meningitis; toxic shock (due to toxin release); and even food poisoning. Most staph are facultative anaerobes, meaning they prefer oxygen. The truly anaerobic species, *S. aureus anaerobius*, rarely cause infection.

This electron micrograph shows round S. aureus *dwelling in the polysaccharide-laden biofilm it has formed on a catheter inserted into a hospital patient.*

(Courtesy of the CDC, Janice Carr.)

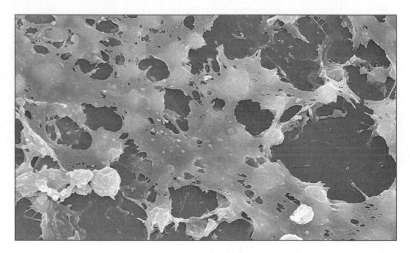

S. aureus usually lives in the nose and armpits, and the usually harmless commensal *S. epidermidis* lives, of course, on the skin. Both of these bugs are the chief pathogens that form biofilms on medical devices, such as joint replacements and implanted heart valves. After being ensconced in biofilms, these bugs become resistant to phagocytosis, which is the chief host weapon against them. *S. aureus* possesses surface proteins that make it easy to get into a host and increase its virulence.

Streptococcus: Commensals and Pathogens

Streptococcus is another genus of round gram-positive bacteria that cause suppurative diseases, although they grow in chains rather than clumps. Members of the genus are often commensals. They can also be important in industry. As commensals, they can be held in check by competition from other microbes and by the normal antimicrobial mechanisms of the innate immune system, such as expulsion by coughing and sneezing, getting trapped in mucus, and phagocytosis. Still, many of the known species are human pathogens that cause respiratory and blood diseases and skin infections. The respiratory infections include not just pneumonia, but also sinusitis and middle ear infections.

Streptococci are often classified in different groups based on their cell wall components. Most human streptococcal disease is the result of Group A streptococci. That includes *S. pyogenes*, a bug that has been a major interest because it is responsible for a great many diseases: pharyngitis (strep throat) and tonsillitis, skin infection (notably impetigo), bone and joint infection, meningitis, and endocarditis.

S. pyogenes is the flesh-eating bacteria of media fame. Despite what you've been told, it's rare. This organism is also of historical importance as the cause of puerperal fever, the often-fatal infection following childbirth. It is one of the more successful pathogens because it can elude phagocytosis. It also evades immune system surveillance because it possesses several antigens that resemble human antigens.

S. pyogenes infections often have serious long-term consequences, such as rheumatic fever and kidney disease. These appear to be the result of immune system actions rather than the direct result of infection. The suspect is a cross-reaction of *S. pyogenes* antigens with host tissue antigens. Rheumatic fever has become less of a health problem in North America, although it is still a great concern in the tropics.

About one in ten people possess *S. pyogenes* in the nose and throat, and more than one in five have *S. pneumoniae*. Infection by *S. pyogenes* and *S. pneumoniae* often come about after normal flora have been disrupted or after infection by a virus.

S. pneumoniae has also been a focus of research attention because it is the chief cause of fatal pneumococcal pneumonia, especially in communities such as nursing homes, and in the developing world. It appears to infect humans only. *S. pneumoniae* is also the chief cause of sinusitis, acute middle ear infection, and conjunctivitis. It possesses a capsule that protects it from phagocytosis. But it's actually a bit fragile in the real world, and person-to-person transmission depends on close contact. A vaccine that is 60 percent effective in preventing pneumococcal pneumonia exists, although it is underused.

Scanning electron micro-graph of S. pneumoniae.

(Courtesy of the CDC, Janice Carr.)

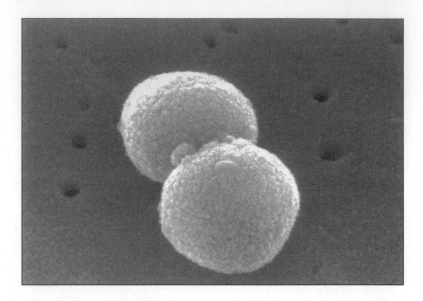

Group B includes *S. agalactiae*, which sometimes causes a fatal disease in infants, including meningitis. Researchers report that up to two in five women carry this bug in their vaginal tracts. It colonizes the majority of newborns, but only 1 to 2 percent develops the disease, probably because their immune system mechanisms, such as phagocytosis, are still immature. Later in infant life, the infection is often acquired in a hospital. Groups B, C, and G streptococci formerly were thought of only as animal pathogens.

Group D, also an animal pathogen, now includes bugs that cause enterococcal infections, the common *Enterococcus faecalis* (which also causes endocarditis) and *E. faecium*. Enterococci are now the second most common human pathogen found in hospitals. These usual residents of the human intestinal tract cause urinary tract and wound infections that are resistant to antibiotics. Water-quality inspectors use this bug to measure the safety of water at swimming beaches.

The viridans streptococci comprise a diverse collection of species that are especially common in the mouth. As you may recall, one member, *S. mutans*, is a major player in dental biofilm and causes tooth decay. Viridans species are suspected of also causing bacterial endocarditis.

Neisseria: Gonorrhea and Meningitis

Neisseria gonorrhoeae, the most notorious bug in this gram-negative group of aerobes, is the cause of a sexually transmitted disease: gonorrhea. *N. gonorrhoeae* can infect mucus membranes in the urethra, cervix, rectum, throat, eyes (especially in newborns, who can become infected during birth and become blind as a result), and then spread to cause, for example, pelvic inflammatory disease. The main symptom is a heavy discharge of pus. This bug, also called the gonococcus, can also spread through the blood to cause skin infections, arthritis, and endocarditis.

> **Little Did You Know**
>
> People are the only known host for *N.gonorrhoeae*, but it is not a commensal. It spreads easily because many of those infected display no symptoms.

In addition to antibodies, host defenses rely in part on the innate immune system. For example, vaginal pH changes during the menstrual cycle make women resistant to infection at certain times of the month. Antimicrobial molecules in urine also appear to offer protection against infection. People with genetic defects in the complement system are particularly at risk for infection. *N. gonorrhoeae* is one of the many bacteria that have developed resistance to multiple antibiotics.

N. meningitidis infection is uncommon, but dangerous. This bug, also known as meningococcus, causes acute meningitis and skin lesions. Both can be fatal. It is the only cause of bacterial meningitis epidemics. In this case, the infection is transmitted by breathing in infective particles.

This bug is present in the nose and throat of as many as one in three people with no symptoms, but less than one percent actually develop the disease. It is believed to strengthen their immunity to serious infection. IgM and IgG appear to be especially important in protecting against disease, but sometimes even a strong antibody response is not strong enough. As is the case with the gonococcus, deficiencies in the complement system increase the risk.

Both bugs are sources of enzymes that attack IgA, an important immunoglobulin in host mucus membranes. The gonococcus, a bacterium that can hijack host cells, forces the cells to engulf them, and uses them as transport into host tissues. This trick is called parasite-directed endocytosis.

It also possesses several variable surface antigens. A gonococcus uses only one of these variants at any one time, which may help it evade immune system surveillance. Other bacteria in this family colonize the mucus membranes of people and other animals, but cause only opportunistic infections.

Bacillus: Anthrax

The term bacillus often means any rod-shaped bacterium, but when spelled with a capital *B* and italicized, it applies only to the genus *Bacillus* (which is, go figure, rod-shaped.) Some 40 species are known. Members of this aerobic gram-positive genus are everywhere and are even used in industrial and pharmaceutical processes. They produce many enzymes and at least two antibiotics, bacitracin and polymyxin. *Bacillus subtilis* is a model organism used in research. Much of what is known about gram-positive bacteria comes from research on this bug.

Bacillus species are mostly saprophytes, which means they eat dead stuff. Although they have often been the culprits in food spoilage, only a couple are normally of medical importance. The potential biological warfare applications of the main one, *Bacillus anthracis*, are discussed in Chapter 17.

B. anthracis is the cause of anthrax. Anthrax has a long history in human medical lore and has even made an appearance in the Bible among the plagues of Egypt. *B. anthracis* became the first bacterium shown to cause disease, by Robert Koch in 1877, and the first bacterial vaccines were against anthrax. However, *B. anthracis* is mostly a pathogen of herbivores, especially cattle and sheep, and apparently cannot be transmitted between people. The animals consume spores along with grass and humans catch it from them, directly or indirectly.

Death of the herbivore can result when the anthrax infection is systemic, which can happen when spores in the gastrointestinal tract germinate and enter the lymph and blood. The bug can keep on multiplying even in a dead carcass. Spores are quite persistent in the environment, especially soil, and can survive tough conditions—the desert, the Arctic—to live again. The bugs themselves are survivors as well. They can manage in a wider range of temperatures, pH, and salt than most other bacteria.

Little Did You Know

B. anthracis got its name from the large black sores it causes; *anthrax* is the Greek word for coal.

Almost all cases of human anthrax affect the skin and come from handling infected material. Eating infected meat can transfer infection to the human intestines, and breathing spores results in pulmonary anthrax.

This is another of the bugs that is wrapped in a capsule that helps it elude phagocytes. There are vaccines for both humans and animals, and both stimulate production of antibodies to anthrax toxin.

Heavy populations of *B. cereus* in improperly stored food—notably rice that has been sitting around unrefrigerated and then not reheated adequately enough to kill all the new spores—causes two severe kinds of food poisoning: vomiting soon after eating an infected meal, and diarrhea that comes on more slowly. This does not result from a toxin that the bug releases, but rather from bacterial enzymes that produce a toxin from the food.

B. cereus can also occasionally cause infections of the blood, endocarditis, meningitis, and other infections. The same conditions sometimes are attributed to other members of this genus: *B. subtilis* and *B. licheniformis*. These infections usually gain a foothold in people with weakened immune systems.

Other members of the genus *Bacillus* are pathogens in insects. This includes notably *B. thuringiensis*, a deadly pathogen of caterpillars. This humble microbe has become a friend to gardeners because the lethal protein it makes has been incorporated into pesticides.

Infections from Our Normal Anaerobic Flora

As you have already seen, some bacterial genera contain mixed populations of aerobes and anaerobes and include both gram-positive and gram-negative species. Some anaerobes that are sometimes classified as rods and sometimes as cocci turn up in as many as a third of clinical specimens, but they have not received much attention from microbiologists. For example, scientists do not know a lot about how these particular anaerobes interact with the humoral immune system.

These particular anaerobes are a diverse group that cause a diverse array of diseases in diverse parts of the body: pus producers such as abscesses (some serious and even lethal), wound infections, and also infections under the skin (cellulitis), blood infections (bacteremia), and pneumonia. Most usually involve several microbes and come about because normal flora overrun damaged tissue. These infections tend to be smelly. In addition to antibiotics, treatment often includes surgical removal of damaged tissue, called debridement.

Brain abscess is probably the most serious clinical outcome; nearly half the patients die. The chief culprits are normal human commensals found largely in the gut: *Bacteroides*, especially *B. fragilis*, and *Fusobacterium*. *Bacteroides* is 20 times more common in the gut than the much better-known *Escherichia coli*, and it is sometimes called the most important clinical anaerobe.

These bugs generally infect the brain by traveling from infection sites elsewhere in the body, such as the ears, lungs, and sinuses. Lung infections from these bugs are often chronic, sometimes fatal, and frequently the result of commensals traveling from the mouth.

Skin and tissue infections from these bugs can be fatal too. Among these are gangrene and the media darling, necrotizing fascitis, the flesh eater. These infections often also include *Staphylococcus aureus* and *Streptococcus pyogenes*. Other genera of clinical interest are *Prevotella* (especially in bacteremia and infections of the female genital tract), *Gemella*, and *Veillonella* (often from human bites!). All of them are among our normal flora.

Clostridium: Gangrene, Tetanus, and Botulism

This genus of spore-forming anaerobes is among the best-studied, disease-causing bugs. We say anaerobes because they are usually classified that way. However, there are also species that can tolerate oxygen, and others that can even grow in it.

Clostridium is also the major exception to the general observation that most anaerobic infections are due to our normal flora. That's because *Clostridium* usually dwells in the soil. However, these bugs are opportunistic and cause an array of disorders: bloody diarrhea, gas gangrene, cellulitis, tetanus, and food poisoning (which this book discusses in Chapter 17). *Clostridium's* chief weapons are the toxins it produces.

Clostridium causes gas gangrene and other wound infections. Gas gangrene is a type of gangrene so called because the *Clostridium* toxin generates gas as it is destroying infected tissues.

Gas gangrene is potentially lethal because of severe shock, a health crisis calling for immediate attention and probable surgery. Neither immune responses nor antibiotics are terribly helpful. The bacterium most commonly responsible is *C. perfringens*, which also causes food poisoning. However, *C. septicum* and *C. novyi* can also cause gas gangrene, and occasionally so can a few other *Clostridium* species.

Toxins produced by two other *Clostridium* species are among the most powerful toxins arrayed against us. They are made by *C. tetani*, which causes tetanus, and *C. botulinum*, which causes botulism. This form of food poisoning is discussed in Chapter 17.

Tetanus, often called lockjaw, is a lethal disease that usually begins as a small puncture wound. Toxin leaves the wound when *C. tetani* cells burst open. The toxin migrates to the central nervous system and causes intense spasms and rigidity that eventually shut down breathing. However, it is no longer common in the United States thanks to routine immunization with tetanus toxoid, a weakened version of the toxin.

The aptly named *C. difficile*, cause of a vicious bloody diarrhea, has become an increasingly important public health problem related to antibiotic resistance. This issue is discussed in Chapter 12.

Mycoplasma

There are more than 100 species of *Mycoplasma*, the smallest bacteria. Because they have few genes compared to other bugs, they make few proteins and so depend on the parasitic lifestyle.

In culture, *Mycoplasma*s form filaments like fungi, which is how the genus got its name. (In Greek, *myces* means fungus and *plasma* means form.) They are parasites on people but also animals, insects, and even plants. For scientists and biotechnologists, *Mycoplasma*s are notable (and notably irritating) because they frequently contaminate cell cultures.

From the standpoint of human health, the most important species is *M. pneumoniae*, a major cause of respiratory infection that affects mostly children. Despite *M. pneumoniae*'s name, the infection does not usually progress to pneumonia and is rarely fatal. This was the first *Mycoplasma* species shown to cause human disease. Other species appear to cause mostly opportunistic infections in the immunocompromised, especially those infected by HIV.

Tiny Tips

If you want to know the minimum equipment for building a successful bug, look at the genus *Mycoplasma*. All you need is a bit of double-stranded DNA containing as few as 500 genes, some ribosomes, and a cell membrane. These are the only prokaryotes without cell walls.

Pseudomonas: Friend and Sometimes Opportunist

There are well over a hundred species of *Pseudomonas*, which are ubiquitous gram-negative aerobes, and sometimes plant pathogens. However, only a few cause human disease, and most of those are opportunistic infections. *Pseudomonas* species cause an estimated 10 percent of nosocomial infections, many of them due to *P. aeruginosa*, whose normal habitats are soil, water, and vegetation.

P. aeruginosa is a free-living rod that is the most heavily studied of the pseudomonads. One expert has declared it to be "the epitome of an opportunistic pathogen" because it hardly ever infects tissues unless they are immunosuppressed or compromised in some way. *P. aeruginosa* can also grow without oxygen and is resistant to many antibiotics. It also resists phagocytosis and the human adaptive immune system with the help of an impervious slime layer it produces.

It produces a strong toxin, known as toxin A, that is similar to diphtheria toxin and appears to block protein synthesis in host cells. It is responsible for a substantial proportion of bacteremia and kills about half of those patients. Death is common among infected patients with cancer and burns.

P. maltophilia, normally found in water and milk, is also an opportunistic pathogen that turns up frequently in clinical studies.

Outside the hospital walls, pseudomonads are mostly friends of humanity and other organisms. They are serious saprophytes—eaters of dead things—and therefore masters of decomposition and biodegradation. They are particularly good at breaking down organic compounds that other microbes can't handle. Eventually they may prove to be a help in handling environmental pollution. The downside is that they are also good at spoiling food.

Brucella: Mostly Animal Diseases

Bacteria of the genus *Brucella* are gram-negative short rods. They are also zoonotics, which means they cause mostly animal diseases (zoonoses), and people mostly catch them from animals. There are six species; only four are human pathogens. They live inside cells and reproduce happily in host macrophages.

In people, brucellosis is like a severe case of the flu (fever, pain, and extreme fatigue), and recovery can take up to a year. People who work directly with animals are most at risk, and the bug can also be transmitted in infected milk that has not been pasteurized.

Brucellosis is an economically important disease of livestock, especially ruminants, as it causes reproductive loss; however, miscarriage is not a feature of the human version of the disease. Animal vaccines exist, and slaughtering the infected animals controls the disease.

Yersinia pestis: Plague

Plague is another human disease transmitted by animals; this one is perhaps the most notorious disease in history. Plague to most people means the Black Death, the pandemic of (probable) bubonic plague in Europe that peaked in the fourteenth century, persisted until the seventeenth century, and killed 200 million people, which was one third of the population. However, that is not the only occurrence of the disease, caused by *Yersinia pestis*.

An outbreak, possibly in the eleventh century B.C.E., is described in the biblical book of Samuel; it was visited on the Philistines for stealing the Ark of the Covenant. Diseases that may have been bubonic plague were described for the next 1,500 years around the Mediterranean, with the first well-established pandemic occurring in the sixth century C.E. It is known as the Plague of Justinian and killed an estimated 40 percent of the population of Constantinople. It then moved on around the Mediterranean to France, and left an estimated 25 million people dead in its wake.

What appears to have been a third pandemic swept through China and India in the nineteenth century. This may actually have been two related diseases, bubonic plague and pneumonic plague, both caused by *Yersinia pestis*. This gram-negative bug is primarily a pathogen of rodents, carried from them to people through fleabites.

> **Little Did You Know**
>
> Bubonic plague gets its name from one of its most prominent symptoms: swollen painful lymph nodes that turn black and are called buboes.

Bubonic plague is extremely contagious and often fatal. As an epidemic persists, the disease changes into the pneumonic form, which has a 100 percent chance of death.

To complete its obnoxious role in human history, *Y. pestis* has also long been used as a biological weapon. In medieval European wars, combatants contaminated water supplies with infected carcasses and catapulted plague victims over the walls of cities under siege. In World War II, the Japanese army bred and released fleas infected with *Y. pestis* and deliberately infected civilians and prisoners of war. Research on this bad bug is in the plan for the U.S. government's bioterrorism preparedness. This subject is discussed more in Chapter 17.

Two other species of *Yersinia* also cause human disease: *Y. enterocolitica* and *Y. pseudotuberculosis*. These create intestinal problems and are spread from animals through infected water.

Antibiotics work against all these bugs, and vaccines are in development.

Haemophilus: Meningitis and Deafness

Despite its name, the bacterium *Haemophilus influenzae* does not cause flu, although it was believed to be the flu bug until the 1930s. Flu is now known, of course, to be a viral disease, and we discussed it in Chapter 14. *H. influenzae* is the most formidable of the *Haemophilus* species, the cause of many diseases. It was the first free-living organism to have its genome sequenced; it's a small genome, with well under 1,800 genes. This bug invades the blood and causes meningitis, especially in young children. Those who survive often suffer from deafness and learning disabilities.

H. influenzae possesses a capsule, which helps it withstand phagocytosis and is believed to contribute to its virulence. There are also strains without capsules. They are less virulent, but can cause inflammation. Other strains of *H. influenzae* cause middle ear infections, bronchitis, pneumonia, and sinusitis. These bugs tend to be opportunistic pathogens that often do not make people sick. There are effective vaccines to treat illnesses that *H. influenzae* cause.

Bordetella: Whooping Cough

The aerobe *Bordetella pertussis* is responsible for pertussis, more commonly known as whooping cough. It is a contagious and often fatal disease in infants, but it is completely preventable with vaccination (the DPT vaccination that also prevents diphtheria and tetanus). There are many millions of cases every year, almost all of them in developing countries. The adult disease is mild. Other members of the *Bordetella* genus cause respiratory disease in animals and birds.

Corynebacterium diptheriae: Diptheria

Corynebacterium diptheriae is another bug that used to be a childhood scourge but succumbed to the DPT vaccine—at least in the developed world, where the formerly dreaded diphtheria has all but disappeared. Its chief weapon is the powerful toxin it produces. It was the first disease found (late in the nineteenth century) to result from a bacterial toxin.

Diphtheria's symptoms include sore throat, high fever, and swelling of the neck that interferes with swallowing. It is contagious and fatal in about 1 in 10 cases. The toxin can lead to congestive heart failure and paralysis. The lesions it causes, classically on the tonsils, throat, and nose, are often covered by a gray-green skin-like structure called a pseudomembrane.

There are two main kinds of diphtheria. One affects the nose and throat (nasopharyngeal), and the other, usually milder, affects the skin (cutaneous). There are harmless versions of this bug, but they can be quickly converted to pathogens by transfer of the toxin gene.

Mycobacterium: Tuberculosis and Leprosy

Human bones that are six thousand years old show signs of it. In the nineteenth century, it was called consumption, and today it is said to be the world's most common infectious disease. Yes, tuberculosis (TB) is still with us. The World Health Organization calls it a global health emergency, with two million dead every year, twice as many as are killed by malaria.

Mycobacterium tuberculosis causes most of it. This bug can survive in the outside world, but can grow only inside a host. It settles in the lower respiratory system, producing a chronic phlegmy cough, fever and sweating, and loss of weight. *M. tuberculosis* can also cause meningitis and infect lymphatics and blood, the genitourinary tract, and the bones.

M. tuberculosis is contagious, but comparatively few of the people that are infected get sick. It is estimated that one third of the world's population—billions of people—are infected and will never know it. TB tends to be a disease of the old, the badly nourished, and the immunocompromised.

The mycobacteria are slow-growing aerobes that possess a lipid-filled waxy cell wall found in this bacterial family only. This tough shell helps these bugs fend off dehydration, extremes of pH, and destruction in host macrophages. However, host T cells can either break open infected macrophages or trigger them to destroy their resident bacteria. Antibodies offer no protection.

> **Little Did You Know**
>
> Resistance to multiple antibiotics has made TB a serious public health problem once again, especially for those infected with HIV. Vaccines exist, but they are not administered routinely.

M. leprae infects the skin and nerves to cause leprosy (Hansen's disease). The disfiguring disease is renowned in history but no longer prominent in medicine, thanks to antibiotics. Other mycobacteria cause respiratory disease that is not TB, along with inflammation of the lymph nodes and infections of skin and soft tissues. People pick up these infections not from each other, but from infected soil and water.

Chlamydia: STD and Blindness

Chlamydia trachomatis causes a common sexually transmitted disease (STD), but it is also the world's most important cause of preventable blindness. That blindness is due to trachoma, a persistent conjunctivitis that leads to scarring of the cornea. The bug also causes a milder conjunctivitis. Several antibiotics cure all three diseases.

Chlamydia are intracellular parasites, and all three known species cause human disease. *C. psittaci* primarily causes a respiratory infection of birds, but they can transmit it to humans and other animals. *C. pneumoniae* infects the human respiratory tract and is transmitted person to person.

Treponema: Syphilis

There is one major pathogenic species in this genus of spirochaetes, which are corkscrew-shaped bacteria that twist to move with the help of flagella between the cell membrane and the cell wall. This pathogen is *Treponema pallidum,* the cause of one of the most notorious STDs: syphilis. It causes genital sores and, if untreated, eventually invades the brain and heart. This process may take decades, but the results are calamitous.

Electron micrograph of spirals of T. pallidum, *the corkscrew-shaped bugs that cause syphilis.*

(Courtesy of the CDC, Dr David Cox.)

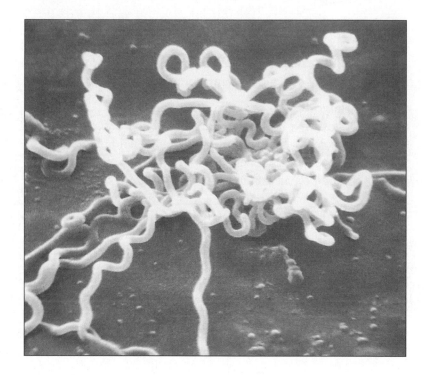

However, *T. pallidum* also causes other diseases with different symptoms. All occur most often in the young. Congenital syphilis is transmitted to an infant during birth. It can cause birth defects and, in about half the cases, death. The diseases yaws (in the tropics), pinta (in Central and South America), and endemic syphilis (in the Middle East) are transmitted through direct contact that is not sexual. All these infections are communicated only from person to person. Penicillin works against them all.

Treponemes are also normal flora in the intestines and mouth, where they are suspected of being involved in gum disease. Treponemes can evade the host immune system because their outer membranes have few surface proteins to set off alarms.

Other Spirochaetes

Borellia burgdorferi is the highest-profile bug in this group of spirochaetes that cause zoonoses. *B. burgdorferi* infects ticks, and ticks bite animals and people, and the result is Lyme disease. Early symptoms include rash, fever, fatigue, muscle and joint aches, and sore throat. Long-term manifestations may be far more serious, although rarely fatal. In addition to arthritis and fatigue, the possibilities include meningitis, numbness and tingling of nerve ends, tremor, memory loss, and even hallucinations.

Other Borellia species that are also transmitted by infected insects cause relapsing fevers with flu-like symptoms; sometimes they can kill the victims. *Leptospira interrogans* causes leptospirosis, which in humans is usually not serious, but can cause organ failure. Leptospirosis affects many animals, an economic problem for the meat and dairy industries. It is transmitted to people not by bites, but by contact with an infected animal's urine.

Legionella: Legionnaires' Disease

Several species of the gram-negative intracellular parasite *Legionella* cause legionellosis (also known as Legionnaires' disease because it was identified after acute pneumonia felled many attendees at a 1976 American Legion convention). The most common species is *L. pneumophilia*, although there are nearly 50 known species, and some authorities think any one of them could precipitate the same symptoms. Sometimes infection results in a flu-like self-limited disease known as Pontiac fever (because it was first recognized in Pontiac, Michigan).

These organisms live in water, including drinking water. After they are in a host, they are among the bugs that commandeer macrophages as residences. Cell-mediated immunity appears critical in dealing with them.

Rickettsiae: Rocky Mountain Spotted Fever

The *Rickettsiae*, which includes bacteria from four genera, live in the cells of insects and mammals and cause several zoonoses. The best known is Rocky Mountain spotted fever, due to species of the genus *Rickettsia* and transmitted via tick bites. It seriously damages blood vessels and is fatal in one out of four cases unless treated with antibiotics. There are several other spotted fevers as well, both in the United States and around the world.

Epidemic typhus, caused by *R. prowazekii* transmitted via louse bites, was an epidemic disease of enormous importance to history. The disease used to be common in jails and during wartime because of the ubiquity of the human body louse in those environments. Millions of Eastern Europeans are said to have died of typhus during World War I. Epidemics are now confined to places with poor hygiene and no access to vaccine. Don't confuse typhus with typhoid fever, which also can be fatal, but is a different disease altogether. Typhoid fever is caused by the bacterium *Salmonella typhi*, which is usually found in contaminated water.

The Least You Need to Know

- Most bacteria do not cause disease.

- Many bacterial pathogens cause disease only in those whose immune systems are weakened by malnutrition or other illness.

- Most bacterial causes of disease have been identified.

- People have developed vaccines, antibiotics, and other ways to combat bacterial pathogens, but the chief weapon is the human immune system.

- Bacteria have several ways of fending off host immune systems.

- Members of several different bacterial families cause human disease.

Pathological Protozoa and Fearsome Fungi: Eukaryote Microbial Pathogens

In This Chapter

- ◆ Protozoa and human disease
- ◆ Fungi and human disease
- ◆ Algae and human disease

You'll recall from Chapter 3 that protozoa, algae, and fungi, in contrast to bacteria and archaea, are eukaryotes. Like bacteria, they cause some human diseases, including a few that are among the most worrisome. In this chapter, we'll tell you about them.

Protozoa and Human Disease

"Protozoa," you'll also recall, is an umbrella term for convenience only, to describe one-celled heterotrophic eukaryotes that are usually mobile and free-living, but not at all related to each other in any evolutionary sense.

In fact, some experts define protozoa by saying what they are not: they are not animals, plants, fungi, or algae.

Many protozoa can infect animals, including people. These infections often cause no symptoms or disease; some of these "infections" are actually helpful, even essential, to our functioning.

Protozoa have remarkable strategies for eluding their host's immune systems. Among them are …

◆ Antigenic masking, in which they pretend to be "self" by covering themselves with host antigens.

◆ Antibody blocking, in which they wrap up in harmless antibodies, which obstruct the antibodies that might do them some damage.

◆ Hiding out in host cells, which (temporarily at least) protects them from immune system surveillance.

◆ Antigen swapping, in which a few protozoa clothe themselves in new proteins that the host immune system hasn't yet learned to recognize.

◆ Immunosuppression, where the protozoan pathogen itself can restrain the host's immune system, slowing down its ability to recognize and attack invaders.

Protozoan Diseases: Malaria

Malaria is the leading cause of death from parasitic disease, and we might even say it's the most important human disease of all.

Malaria is not on the radar screens of most of us in the developed world, but its toll is horrendous: an estimated 500 million cases every year, mostly in the tropics, and well over a million deaths, mostly in sub-Saharan Africa. The disease disproportionately affects kids under five and pregnant women. Every minute of every day, at least one small child dies of malaria.

Malaria: The Basics

Malaria symptoms include high fever, chills, joint pain, vomiting, anemia, and convulsions. Most of these are due to toxins in by-products of the pathogen's metabolism. The severest forms of malaria result in anemia, kidney failure, and brain disease that can be mentally handicapping for life when the disease doesn't kill the patient first.

The malaria disease organism is a protozoan parasite in the group known as sporozoans, which means it lacks a means of getting around by itself. Transport is obligingly provided by the mosquito *Anopheles gambiae*.

There are more than 160 species in the malaria-causing protozoan genus *Plasmodium*. Many infect rodents and other animals, including birds and reptiles. Only half a dozen infect humans, with four species accounting for almost all malaria. Eight out of ten cases, and nine out of ten deaths, are due to *P. falciparum*.

The mind-boggling details of the parasite's intricate life cycle have been known for well over a century:

- *Plasmodium* cells called sporozites, carried by a female mosquito, are injected into the human host as she feasts on host blood.

- The sporozites travel to the liver, where each parasite makes thousands of copies of itself.

- The copies burst from the liver (an event that causes the characteristic malarial fever) and invade red blood cells.

- The parasites in red cells also make copies that leave to penetrate new red blood cells.

- Some parasite cells in the blood differentiate into sexual pre-reproductive forms, which are then slurped up by another mosquito along with her human blood meal.

- Those sexual forms settle down in the mosquito's gut, fertilize each other, and then bore through the gut and attach themselves to its outer membrane.

- These cells divide and develop into thousands of sporozites.

- The sporozites move on to the mosquito's salivary glands, where they await injection into a new host when next the mosquito dines on human blood.

Malaria and the Immune System

Malaria parasites are among those organisms skilled at evading the human immune system by hiding out in liver and blood cells. Researchers have recently described details of this evasion in *P. falciparum*, which employs a succession of cloaking devices—camouflage proteins—to make itself invisible.

When the parasites invade a host's red blood cells, they adorn each occupied red cell with a special protein that helps prevent a host immune system from recognizing the invaded cell as nonself. Eventually the immune system penetrates the disguise and starts making antibodies, but by that time some members of each new parasite generation have draped themselves in a different cloaking protein that once more can fool the immune system. For a while, at least, parasites dressed in the new protein cloaks don't appear to be invaders and can infect new red cells. Researchers estimate that *P. falciparum* has some 50 genes that can code for these cloaking proteins, but only one is turned on at a time.

A photomicrograph of a smear of blood containing red blood cells that have been infected by malaria parasites, which appear as small rings inside the cells.

(Courtesy of the CDC, Dr. Mae Melvin.)

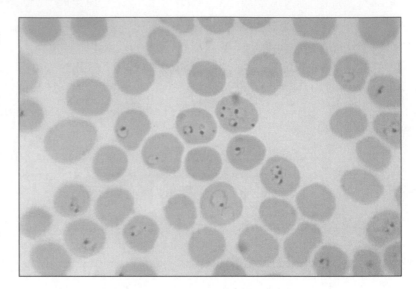

Tiny Tips

The better example of genetic resistance to malaria is more recently discovered and not so well known as hemoglobin S. It is hemoglobin C. This variant can result in mild anemia, but mostly it is symptom-free while offering excellent protection against malaria.

The malaria parasite has shaped human evolution by killing people who are most susceptible to it, permitting those who can resist to survive and produce descendants. This resistance appears to involve several genes, including some still unknown immune system genes. However, other such genes were identified beginning more than 50 years ago.

For instance, people with a single copy of certain genes for red blood cell variants are almost entirely resistant to malaria. These variants, which seem to block the parasite's entry into red cells, are particularly common in Africa, where malaria is particularly common too.

The best-known example is the gene for hemoglobin S. Unfortunately, a child who inherits two such genes, one from each parent, will develop a serious, often fatal, disease: sickle-cell anemia. But the parents, each with just a single copy of the gene, will remain healthy. They can produce several children with only one copy of the hemoglobin S gene, who will remain healthy, too.

Preventing and Curing Malaria

Most malaria prevention employs the same anti-mosquito strategies as with the arbovirus diseases: bed nets, long sleeves and pants, and insecticide. Eliminating breeding places, draining swamps, getting rid of standing water, and encouraging basic hygiene has worked in many parts of the world, such as the southern United States. A good deal of this success was due to the powerful insecticide DDT. Unfortunately, it was so powerful that it caused environmental pollution and killed many organisms besides the *Anopheles* mosquito. DDT use is now discouraged in developed countries.

Treatments for malaria can work well if given early enough. There are several drugs that can treat the disease and even prevent attacks. Unfortunately the parasite is becoming resistant to the most important ones, such as chloroquine, resulting in malaria increase in some locales. Even more unfortunately, even effective anti-malaria measures can be too costly in some parts of the world.

P. falciparum's genome has been sequenced, but drugs that might emerge from studies of its genes are many years away. The mosquito genome has been sequenced, too. The hope here is to create a malaria-resistant mosquito via genetic manipulation. This talk has raised an ironic question: if mosquitoes can resist malaria parasites, will they be healthier and more able to transmit other possibly deadly diseases?

Vaccines are of course in development and some early clinical trials report success. But production and widespread administration is several years away at best.

Protozoan Diseases: Gastroenteritis, a.k.a. Diarrheal Disorders

The World Health Organization publishes an annual survey listing the most important causes of death. It reports that diarrheal diseases cause 17 percent of deaths in children under five. That is nearly one out of five deaths of small children attributable to disorders that are largely preventable.

Total deaths, adults as well as kids, are estimated at two million annually. The death count in the United States is only 6,000 per year, but hospitalizations number nearly a million, and U.S. residents each average between one and two acute episodes annually. The world total number of annual cases is estimated at three to five billion. Yes, that's *billion* with a *b*.

"Diarrheal diseases" is another umbrella term that covers many different causes of gastroenteritis, the medical term for inflammation of the lining of the stomach and gut leading to diarrhea. We've already talked about several of the responsible microbes: viruses, especially the notorious rotavirus, and bacteria like *Salmonella* and *Shigella*. We'll discuss the topic again in connection with food in Chapter 17.

But a large proportion of the 17 percent of childhood deaths, and about 15 percent of the more ordinary GI problems suffered occasionally by us all, are due to protozoa.

Entamoeba histolytica

Amoebas (sometimes spelled *amebas*) are often animal parasites but not many of them colonize humans and even fewer are pathogens. At least half a dozen species regularly live in the human gut but never cause disease.

Much less common is the major human pathogen, *Entamoeba histolytica*, endemic in the tropics but not terribly common in the developed world. In nine out of ten cases, there may be no symptoms or only mild ones. But, in that one case out of ten, it causes amebiasis or amebic dysentery. That's an exceptionally nasty diarrhea accompanied by blood and pus, plus fever and severe abdominal cramps. Sometimes *E. histolytica* also burrows into the bloodstream and travels to the liver and other organs. It is said to be the third leading parasitic cause of death.

Photomicrograph of the durable cysts of Entamoeba histolytica, *which can survive in the environment for a long time.*

(Courtesy of the CDC.)

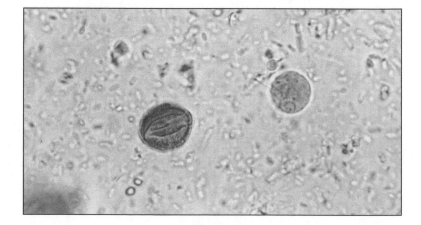

These microorganisms (in the form of cysts) are transmitted by the fecal-oral route of course, generally in contaminated food or water, but also by direct contact, including sexual contact. Once in the gut, the cysts transform into an active feeding stage organism that divides into eight new amoebas. They encyst while traveling through the colon and are shed into the environment as durable cysts that can be ingested by other organisms eventually. This infection can be treated with antibiotics, but often recurs.

Giardia lamblia

We have met this formerly "harmless commensal" before, in Chapter 9. It is a common cause of diarrhea all over the world and the most common nonbacterial cause of diarrhea in North America, found in both drinking water and bodies of water used for recreation. It infects up to 40 percent of the population in developing countries. However, many of those infected display no symptoms. *Giardia* slides by the immune system with the help of surface proteins that disguise it, so it tends to recur. It is only rarely fatal. It also infects cats.

> **Little Did You Know**
>
> *G.lamblia* is also the leading cause of "beaver fever," which usually afflicts people while they are on camping trips where sanitized water may be difficult to find.

Cryptosporidium

This protozoan was not recognized as a human pathogen until 1976. The disease it causes, cryptosporidiosis, consists mostly of diarrhea and sometimes respiratory symptoms. It is growing more common because it particularly attacks those with weakened immune systems, such as AIDS patients. Like *Giardia* it is passed on via contaminated water. It also infects many kinds of animals and can be transmitted to people through livestock or pets. This infection is exceptionally difficult to treat. Three other less common spore-forming protozoans cause similar disease in similar ways. All are rarely fatal.

Protozoan Diseases: *Trichomonas vaginalis*

Trichomoniasis—commonly called just trich—is probably the most common sexually transmitted disease on earth. Worldwide prevalence is estimated at 170 million cases annually (more than gonorrhea or Chlamydia). It is transmitted almost exclusively by genital to genital contact, and occasionally from mother to baby during childbirth.

T. vaginalis, the flagellated protozoan responsible, infects both sexes, but men almost never have symptoms and so the number infected is unknown. Infected women usually produce a smelly vaginal discharge and report maddening vaginal itching and burning. There are about three to five million cases in the United States every year. The infection rarely produces complications, is easily cured with anti-protozoan agents like metroniolazol, and is never fatal.

Protozoan Diseases: More Ailments Carried by Insects

The trypanosomes, protozoans with a single flagella, infect insects, and a few also infect secondary hosts. These secondary hosts may be plants or vertebrates. Some of them are us; the infecting protozoans cause different kinds of trypanosomiasis, among the most serious human diseases on the planet, and also several serious animal diseases.

Trypanosome's strategy for evading the host immune system is similar to the one we described for malaria. Once in a host, a few of them switch their glycoprotein coats. This makes them invisible for awhile and enables them to infect new cells until the immune system begins to recognize and attack the new coating. Trypanosomes possess as many as 1,000 different coat genes, which they employ one at a time.

Trypanosoma cruzi

American trypanosomiasis, more often called Chagas's disease (after its early twentieth-century describer, physician Carlos Chagas) can be an acute flulike illness shortly after infection, especially in the young. Although many of those infected are symptom-free, the infection lasts forever.

Trypanosoma cruzi works its damage on the heart and GI tract for years after infection. Chagas's disease is the leading cause of congestive heart failure in areas of Latin America where it is endemic, and it also damages the esophagus and the colon. About 90 million people are at risk. The annual death toll is 70,000.

As is the case with many infections, occasionally the trypanosomiasis parasite is transmitted through blood transfusion, pregnancy, and breast milk. But the usual method, insect-based transmission, is uniquely peculiar and revolting.

The protozoan is transmitted by several different large insects known locally by many names, among them assassin bug, benchuca, vinchuca, kissing bug, chipo, and barbeiro. The insects pick up their infection by biting infected mammals and come out at night to bite people while they sleep. But infection is not transferred by the act of biting itself, as it is with mosquitoes and malaria.

Instead, the parasites lurk in the insect's feces. The insect defecates while biting, usually on the sleeper's face. The parasite moves in when the human victim inadvertently rubs the infected insect feces into the bite or other wound.

Little Did You Know

The British naturalist Charles Darwin, who described the process of evolution by natural selection, began to formulate his ideas while studying plants and animals on his years-long trip to South America in the 1830s. After he returned from that trip, mystery illnesses incapacitated Darwin for much of the rest of his life. One theory is that he was suffering from Chagas's disease. In the detailed diary he kept of his voyage, he described being bitten by vinchuca, "the great black bug of the Pampas." This insect is one of the known vectors of *T. cruzi*.

The protozoan infects an estimated 16 to 18 million people in the Western hemisphere, and there are about 200,000 cases of Chagas's disease every year. It is especially prevalent in Mexico and Central and South America. The disease is rare in the United States, although many thousands of immigrants are infected.

Trypanosoma brucei

African sleeping sickness almost disappeared in the 1960s, thanks to a sustained surveillance program set up after three epidemics had swept through the continent in the preceding decades. But the disease has made a strong comeback as African healthcare systems and surveillance collapsed under the chaos of civil war and massive population movements.

The World Health Organization (WHO) estimates that 300,000 to 500,000 people have the disease. In some villages, the rate approaches 50 percent, and sleeping sickness causes more deaths than AIDS.

African sleeping sickness, often known as human African trypanosomiasis, or HAT, is the result of an infection by *Trypanosoma brucei*. It is transmitted—more conventionally and less colorfully than Chagas's disease—in the bite of an infected tsetse fly, a large fly common in sub-Saharan Africa.

There are two forms of HAT, caused by two nearly identical trypanosomes. *T. brucei gambiense*, found in central and western Africa, causes a chronic disease. *T. brucei rhodesiense*, found in southern and eastern Africa, causes an acute response soon after infection and is more virulent. Animals also carry these parasites, but they commonly are infected by different *T. brucei* species. Trypanosomiasis is a serious problem in cattle.

If caught early, the disease is curable, which is why consistent screening of patients is crucial. Early symptoms include fever, headache, and joint pain. In the later, neurological stages, the parasite invades the brain, bringing on confusion, violent behavior, convulsions, and sensory disturbances.

Late-stage HAT is notable for disrupting daily biological rhythms, which is how the disease came to be called sleeping sickness. Victims suffer nighttime insomnia and are overcome by sleep during the day. Treatment at this stage can prevent death, but not the characteristic behavioral disturbances. Without treatment, patients lapse into a coma and die.

Leishmania

Leishmania comprise many species of trypanosome parasites residing in the tropics worldwide, except that they don't seem to have infected the South Pacific and Australia yet. These protozoans are transmitted to people and animals in the bites of infected sand flies. Sand flies are extremely tiny insects sometimes called sand fleas or no-see-ums. They pick up the parasites by ingesting infected macrophages from humans or animals during a blood meal.

Some 20 *Leishmania* species infect humans as well as animals, and others infect animals alone. The disease they cause is known, not surprisingly, as leishmaniasis, but it has a number of different manifestations.

The most severe human form, visceral leishmaniasis (also known as *kala azar*, Hindi for "black fever") results in fever, greatly enlarged spleen, wasting and weakness, and darkening of the skin. Nine out of ten cases occur in just five countries: Bangladesh, India, Nepal, Sudan, and, jumping to the other hemisphere, Brazil.

Early treatment is 90 percent effective, but if the disease is not treated, the death rate approaches one in four. Effective treatments date from the 1930s and are still in wide use, but the parasites are developing resistance. There is no vaccine. People who have been treated successfully are almost completely protected against reinfection, but only against the same *Leishmania* species.

The most common form, rarely fatal, is cutaneous leishmaniasis, which leaves sometimes-deforming sores all over the body. The mucocutaneous form can lead to disfiguring damage to the nose and mouth.

A ten-year epidemic that began in Sudan in the 1980s reportedly killed 100,000 people, a third of the population in affected areas. The Centers for Disease Control and Prevention (CDC) estimates that there are 1.5 million new cases of cutaneous leishmaniasis every year, along with half a million new cases of visceral leishmaniasis.

Protozoan Diseases: *Toxoplasma gondii*

There's a good chance you've been infected with this ubiquitous parasite, one of the most common in the world. Up to 40 percent of U.S. residents have antibodies against *T. gondii*, and the rate in Germany and France is twice that. At least eight out of ten infections are symptom-free, or end in a mild illness somewhat like infectious mononucleosis. But the parasite can cause a serious disease.

Fetal toxoplasmosis can end in miscarriage or stillbirth, and about 8 percent of infected newborns either die or go on to severe brain abnormalities, seizures, mental retardation, and deafness. Toxoplasmosis is common in people with weakened immune systems where it most often affects the brain or the lungs. In these patients, untreated encephalitis is nearly always fatal.

T. gondii infects many kinds of birds and animals, but its favorite host is cats, both domestic and wild, who eat infected animals or raw meat. In fact, the parasite can pass through its sexual reproductive stage only in the cat gut, after which (encased in a tough covering and now called an oocyst), it is dispatched into the world in cat feces.

Little Did You Know
T. gondii often settles in the eye. It is the single most common cause of eye inflammation that can end in blindness.

Infected cats are symptom-free but can shed millions of oocysts a day for weeks, and each oocyst can survive in a moist environment for up to a year. A new host becomes infected by ingesting oocysts in cat litter or contaminated soil or water. The parasite can also be transmitted from cysts buried in the tissues of food animals if the meat is not thoroughly cooked; infected lamb, beef, and pork are common in many parts of the world. And an infected pregnant woman can pass the parasite to her fetus through the placenta.

The Parasitical Puppeteers: Can Parasites Control Your Behavior?

Here is one of the creepiest characteristics of protozoan parasite infections: there is increasing evidence that parasites mold the behavior of their hosts to suit the parasites' purposes—especially the purposes of getting to new hosts and making more parasites.

We talked about this phenomenon in Chapter 9 when we discussed the many ways the bacterium *Wolbachia* manipulates its insect hosts. Protozoan parasites, it turns out, are also skilled parasitical puppeteers that stage-manage their unwitting hosts' behavior.

One example: cats are the ultimate hosts of *T. gondii*, but one of the intermediate hosts is the rat. The parasite infects cats after they have eaten infected rats. Researchers have shown that infected rats appear to lose their fear of cats, thus increasing the chance that they will be caught and eaten, allowing the parasite to transfer from rat to cat.

This finding seems to be widely accepted, but studies trying to link *T. gondii* infection with behavior in its human hosts are far more controversial. Researchers have reported that infected people have longer reaction times and are more likely to have traffic accidents than the uninfected. Also, infection appears to correlate with a personality trait called novelty-seeking, thought to be somewhat lower in infected people.

These human studies have not been widely accepted, and in any case it's not clear how such behaviors could increase *T. gondii*'s chance of infecting new hosts. But in view of all the other data suggesting that parasites do shape host behavior, it is likely researchers will keep looking.

The malaria parasite is another example of parasite puppeteering, in this case shaping the behavior of its mosquito host. Researchers have so far unearthed two examples of this tactic.

They report that mosquitoes that have ingested parasite-laden blood are at first more cautious about looking for another meal than uninfected mosquitoes. This caution helps the mosquito survive long enough for the parasite to go through its reproductive cycle in the mosquito gut. When the parasite is ready to seek new hosts, its current host's behavior changes again, and the mosquito becomes more aggressive in seeking blood. Infected mosquitoes are more likely than uninfected mosquitoes to bite more than one person in the course of a night, and they spend longer feeding on each victim. That helps increase the chance that parasites will be transferred to new human hosts.

In the other example, researchers report that the malaria parasite makes its human hosts smell particularly delicious to a passing mosquito in search of a meal. That presumably increases the parasite's chances of entering a new mosquito host. After people have been treated successfully for malaria infection, they are no longer especially attractive to mosquitoes.

Fungi and Disease

As we observed in Chapter 3, fewer than 100,000 species of fungus have been identified, but scientists suspect that a million or more kinds of fungi exist in the world. Still, only about 400 of these are known to cause human disease. In people who are healthy to begin with, *mycoses* are uncommon, and most of those that do occur are mild.

Skin infections are commonly fungal. These are annoying, unsightly, and persistent, but not usually dangerous. A few fungi can be very dangerous. Given the increased number of people with weakened immune systems—from organ transplants and diseases of the immune system, especially HIV infection—opportunistic fungal infections are a growing medical problem. Fungi also pose economic troubles, since they cause disease in a great many plants and animals that people depend on.

def•i•ni•tion

Diseases caused by fungi are known as **mycoses**.

Fungal Diseases: *Pneumocystis jiroveci*, formerly *P. carinii*

P. jiroveci (pronounced yee-row-VET-zee) is the perfect organism to serve as the transition from talking about human diseases caused by protozoa to those caused by fungi. That's because it is a yeast-like fungus that until 1988 was classified as a protozoan. It also used to be called *P. carinii* but the human form was renamed in 1999 for the scientist Otto Jirovec, who first described it in humans. Many other *Pneumocystis* species infect other animals.

P. jiroveci causes pneumonia, but is not a problem except for people with weakened immune systems, especially AIDS patients. It is a classic example of opportunistic infection. Indeed, a big rise in the number of cases of *Pneumocystis* pneumonia (called PCP) early in the 1980s first gave U.S. physicians a clue that a vicious new disease, the one that turned out to be AIDS, had appeared and was turning into an epidemic.

In the early days, PCP affected three out of four AIDS patients. The rate has gone down since good AIDS treatments have become more common; incidence in treated populations is between 10 and 20 percent. PCP is still a major cause of death among the untreated. Somewhat surprisingly, the PCP rate is fairly low (around 9 percent) among AIDS patients in the developing world. That is thought to be due either to PCP being underdiagnosed or to AIDS patients succumbing to other respiratory diseases, notably tuberculosis, first.

Fungal Diseases: *Candida* and Candidiasis

Candida is the quintessential opportunistic pathogen. It has exploited new technology and new patterns of human mobility to greatly extend its reach. The range of diseases these fungi cause is enormous. They include annoying-but-not-serious infections of the skin and mucous membranes and invasive, life-threatening disorders that potentially spread to all organ systems and every part of the body.

These invasive infections are the result of increased penetration into the blood circulation and deep tissues. Death rates range up to 40 percent. Penetration is often aided by medical technology such as catheters. These breach the body's first line of defense against fungal pathogens: the usually impenetrable barriers of skin and mucous membranes.

As is so often the case with the few fungal human pathogens, *Candida* produces serious disease, sometimes virulent, especially in people whose immune system functions are suppressed (as in organ transplantation) or injured (as in HIV infection.) As the numbers of these people increase, so does candidiasis.

Wee Warnings

Candida is a normal human commensal, but candidiasis is now also the leading human fungal infection. This fungus is almost as significant a source of nosocomial infections as bacteria are.

The *Candida* genus of small yeasts comprises at least 150 species, but fewer than a dozen cause human disease. The most troublesome species is *Candida albicans*, responsible for well over half the infections. Doctors don't usually see the others unless *C. albicans* got there first. *C. glabrata* accounts for up to 20 percent of infections, a percentage that is growing in part because this fungus is less susceptible to standard treatments than the others.

Some forms of candidiasis, sometimes called thrush, are pretty common even in people whose immune systems are in good shape. In the United States, one in three infants gets oral thrush, and the rate is even higher in the developing world. It takes the form of small sores in the mouth that enlarge to white patches. Infant thrush is usually mild and goes away by itself. The infection can also show up as diaper rash.

Rampant opportunistic oral candidiasis in an HIV-infected patient.

(Courtesy of Sol Silverman, Jr., DDS.)

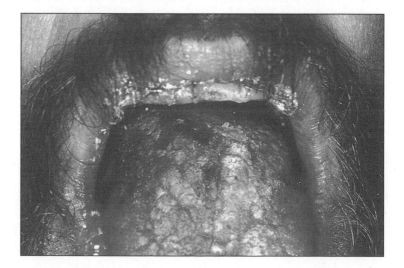

We discussed vaginal candidiasis briefly in Chapter 3. Medical folks call it vulvovaginal candidiasis, but it is more commonly known as a vaginal yeast infection and three out of four women are intimately familiar with it and its maddening itch. It can be an STD (men can get a usually asymptomatic form and pass it on), but also can be the result of hormonal changes and antibiotics. As we have said, the yeast is a normal commensal, but can overgrow wildly when body chemistry changes or competing microbes are killed off.

Fungal Diseases: Allergies, Asthma, and Mycotoxins

We discussed allergies briefly in Chapter 10. Nobody is certain how many people suffer from allergies to fungi (chiefly molds), but the number is huge. Death from mold-induced anaphylactic shock is rare, but illness is common. And mold may be far more dangerous in asthma patients.

Symptoms include runny nose, eye inflammation, coughing, wheezing, sore throat, sinusitis, and sometimes flulike symptoms. In contrast with most pathogens, dead mold can produce these allergic reactions just as well as live mold. The cure, therefore, involves getting rid of mold remains, not just disinfection. Fungi known to produce allergic and asthmatic reactions, often seasonally, include members of *Cephalosporium*, *Alternaria*, *Aspergillus*, and our old friend *Candida*.

Fungi can also produce toxins that make people sick, but this is almost always due to eating the toxin, not just breathing it in. These mycotoxins typically are produced by fungi that have infected a crop plant. Examples include aflatoxin produced by *Aspergillus* species, ochratoxin A produced by *Penicillium* and *Aspergillus*, and a number of toxins from *Fusarium*, some of which cause severe illness and death.

Only some fungi can produce mycotoxins. Moreover, not all mycotoxins are harmful, at least to people. Fungi make toxins to defend themselves against their competitors, which are mostly other microbes. That includes other fungi, and our own species has exploited that weapon. Medical personnel use *Penicillium* toxins to treat dermatophyte infections.

Fungal Diseases: Ringworm and Other Dermafflictions

We put this group of skin disorders next not because they are grave and scary but because they are close to universal and often deeply embarrassing. Few people have escaped one or more of these fungal invasions, which include athlete's foot, jock itch, nail infections, and that memorable humiliation of many a schoolchild, ringworm of the scalp.

All these and more are due not to worms at all, but to a group of fungi collectively known as dermatophytes. The infections are called, collectively, dermatophytosis or tinea seperately. Thus tinea barbae when it occurs in the beard area and neck, tinea capitis for scalp ringworm, tinea corporis on the trunk and extremities, tinea pedis is athlete's foot, tinea cruris is jock itch, and so forth.

These fungi are very contagious and afflict the healthy as well as the immunocompromised. They are transmitted through direct contact and, quite often, from surfaces of everyday objects: combs, hats, bedding, floors, bathtubs, and, famously, locker rooms and hot tubs. The best prevention, as we keep preaching, is ordinary hygiene. There are effective treatments for most of the infections, many available without prescription. The major exception is nail fungus, which is very hard to get rid of.

Tiny Tips

Nail fungus is usually called onychomycosis. It can be due to other fungi as well as dermatophytes.

Fungal Diseases: Aspergillosis

This is a diverse group of mostly opportunistic diseases caused by members of the genus *Aspergillus*, ubiquitous molds which infect mostly the sinuses and lungs.

There are four main diseases:

- Allergic bronchopulmonary aspergillosis (ABPA) is an allergic reaction to *A. fumigatus* and happens mostly in people with asthma or cystic fibrosis.

- Chronic necrotizing *Aspergillus* pneumonia.

- Aspergilloma is a ball of fungus that grows in an existing lung cavity.

- Invasive aspergillosis is a consequence of severe immune system malfunction, as in AIDS, and is often fatal.

Fungal Diseases: Blastomycosis

The organism responsible for blastomycosis is *Blastomyces dermatitidis*, found in soil. Inhaled as spores, the microbe develops into yeast in the lungs and may spread throughout the body, especially to the skin. Skin lesions begin as pustules and become ulcers. About half the infections are asymptomatic.

Blastomycosis is a flulike or pneumonia-like ailment, endemic in the southeastern and central United States. It affects many animals too, especially dogs, and is found elsewhere in the world. It is rarely fatal in the healthy, but in AIDS patients the death rate can be 40 percent.

A Few Other Fungal Diseases

Coccidioidomycosis is also known as cocci, desert fever, and San Joaquin fever. It begins as an acute respiratory infection caused by inhaling spores of *Coccidioides immitis*. That's usually where it ends, and half of those infected display no symptoms at all. In those with other health problems, it can go on to pneumonia, chronic lung disorders, troubles of the skin, bones, and joints, and, the most worrisome complication, meningitis. The infection is endemic in the southwestern United States and Central and South America.

Histoplasmosis is due to *Histoplasma capsulatum*. The soil fungus is found all over the world but is endemic to the U.S. Midwestern river valleys. It likes soil enriched with bird and bat droppings. *H. capsulatum* infects an estimated 250,000 people in the United States annually, but less than 5 percent of them get sick, and most of these have weakened immune systems. It mostly affects the lungs. A related fungus, *Histoplasma duboisii*, is responsible for African histoplasmosis, which mostly affects skin and bones.

When we tell you that sporotrichosis is also known as peat moss disease and drunken gardener's disease, you will not be surprised to learn that it is caused by another soil fungus, *Sporothrix schenckii*. It occurs mostly in the American tropics, resulting from direct contact rather than inhalation. The fungus spreads through the body through lymph channels and mostly causes pus-filled skin lesions. It's a slow-moving disease and it can take awhile to cure.

Algal Diseases

Algae are eukaryotes, too, but on the few occasions when algae pose a threat to human health, they don't do it via infection. Instead they do it indirectly, for example by polluting waters, which we discussed in Chapter 3. Algae also can be a source of food poisoning, especially shellfish poisoning, a topic we'll take up in Chapter 17.

The Least You Need to Know

◆ Many protozoa infect people and animals without causing disease and they have several strategies for eluding the host immune system.

◆ Malaria, which may be the most important human disease of all right now, is the leading cause of death from parasitic disease.

◆ Diarrheal diseases cause a significant proportion of deaths in children under five, and a significant proportion of those deaths are due to protozoa, among them *Entamoeba histolytica, Giardia lamblia, Cryptosporidium.*

◆ There is evidence that some parasites, including *T. gondii* and *P. falciparum*, can control the behavior of their nonhuman hosts in order to increase the parasites' chances of infecting new hosts; it is not yet clear if people's behavior can be shaped by parasitical puppeteers.

◆ Fungal infections normally have a hard time penetrating the body's outer barriers of skin and mucous membranes, so they tend to be serious problems mostly for people with severely weakened immune systems or who have undergone medical procedures that breach those barriers.

17

Food Diseases and Emerging Diseases

In This Chapter

- ◆ When the simple act of eating causes infectious disease
- ◆ Spoiled food may not make you sick
- ◆ Keeping food safe to eat
- ◆ Humans aren't the only victims of diseases
- ◆ Emerging microbial diseases
- ◆ Do microbes cause chronic diseases, too?

In this chapter, we delve more deeply into topics introduced briefly in previous chapters. These topics include infectious diseases where the "vector" is food. Also related to food is information on how microbes spoil foods and how to prevent such spoilage. We also take up plant and animal diseases very briefly.

In this chapter, too, you will find a topic of immense interest to public health authorities around the world: emerging diseases, which can be either brand-new ailments or ones that have been around for a long time but are now assuming new prominence. Finally, we discuss the growing evidence that microbes may play a role in diseases we have not traditionally thought of as infectious, such as heart disease and diabetes.

Foodborne Disease and Food Poisoning

The term "food poisoning" usually means an episode of gastroenteritis that comes on shortly after eating food containing specific pathogens or the toxins they produce. Symptoms are abdominal pain, vomiting, diarrhea, and often complete collapse. The results of food poisoning can be devastating while they are occurring, but they almost always disappear quickly, sometimes in a day.

The worst consequence of most food poisoning is dehydration—loss of fluids and essential electrolytes. If dehydration is treated promptly, food poisoning is hardly ever serious except in people already debilitated by other diseases or old age.

Not all foodborne disease is food poisoning, though. People can get the prion disorder called variant Creutzfeldt-Jakob disease, for example, by eating beef from animals with bovine spongiform encephalopathy (mad-cow disease), which contains prions. But eating contaminated beef causes no acute illness; the lethal symptoms don't appear until years after the fateful meal. We discussed prion diseases in Chapter 4. Another special case is ingestion of toxins from dinoflagellates, which people usually acquire by eating shellfish infested with these algae; we discussed this in Chapter 3.

The Centers for Disease Control and Prevention (CDC) estimates that food poisoning is responsible for 76 million illnesses every year in the United States, including 325,000 hospitalizations and 5,000 deaths. In the overwhelming majority—62 million cases; 265,000 hospitalizations; and 3,200 deaths—the pathogen responsible is unknown. But when food poisoning pathogens are identified, they turn out to be bacteria in two out of three cases.

Tiny Tips

In these globetrotting times, diarrhea is the most common ailment among travelers. The microbe most often identified in these unpleasant encounters is the bacterium *Escherichia coli*. Diarrhea from food poisoning is often brief and usually lasts less than two weeks. If it goes on for longer, the cause is probably not food poisoning.

The CDC reports that nearly all food poisoning is the result of improper food handling, most frequently letting food remain at temperatures that disease microbes love and grow in happily. Other common causes include insufficient cooking, contamination from other infected foods, and infected food handlers. Four out of five cases of food poisoning come from food prepared by commercial or institutional sources, not Mom's home cooking.

The organisms that cause these diseases are all transmitted by ingesting infected feces, usually in contaminated food or water. These feces come from a huge range of species, everything from pet iguanas to other people.

The CDC lists the following among the microbes most commonly identified in food poisoning:

- *Campylobacter* from undercooked or raw chicken

- *Salmonella*

- *E. coli* 0157:H7 from food or water contaminated with cattle feces; this bug is behind large outbreaks of food poisoning from fast-food hamburgers

- Norwalk viruses or noroviruses (discussed in Chapter 14)

Campylobacter

Campylobacter are among the most common infectious organisms that plague humanity. Several species cause human disease, but the most important is *C. jejuni*. In addition to attacking parts of the gastrointestinal (GI) tract directly, some strains also produce a toxin and generate a bloody diarrhea. Death is rare.

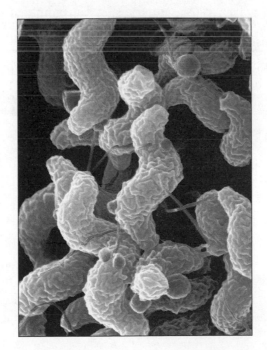

Scanning electron microscope image of Campylobacter jejuni, *the chief cause of food poisoning in the United States.*

(Courtesy of De Wood.)

These infections tend to recur in those with weakened immune systems. Chickens are believed to be the source of well over half of *Campylobacter*-related disease. These bugs are also a big factor in traveler's diarrhea, especially in southeast Asia.

Salmonella

There are thousands of subspecies of *Salmonella*, but not many cause human disease. *Salmonella typhi*, for example, causes typhoid fever, but this terrible disease is not usually classified as *Salmonella* food poisoning, which is called salmonellosis. Typhoid is transmitted via contaminated food or water. That makes it a foodborne disease, but not food poisoning. Internationally there are an estimated 12–33 million cases of typhoid every year, but no one has yet tackled estimates of what must be many times that number of just plain salmonellosis.

In addition to infecting chickens, *Salmonella* infects their eggs as well as nonpoultry, like beef and produce. Safety measures in the industry have now reduced these hazards although thorough cooking is still recommended. Sometimes that isn't possible; recent outbreaks have been due to melons that were not washed thoroughly before being placed on a salad bar. This bug is also a major cause of traveler's diarrhea.

Salmonella that make it through the stomach acid attach themselves to particular cells of the intestinal lining. They cause symptoms by damaging that lining directly, but also wait to be gobbled up by phagocytes. There they reproduce and burst forth, precipitating a flood of body fluids and electrolytes; researchers don't yet understand all details of this mechanism. These bugs are often drug-resistant but death is rare.

Little Did You Know

Bugs we have discussed elsewhere also belong in this chapter. Among its other bad deeds, the ubiquitous *Staphylococcus aureus* also causes a special kind of food poisoning. In this case, the symptoms are due not to the bug itself, but to an enterotoxin it produces. *Listeria monoctogenes* causes miscarriages. It is regulated by the Food and Drug Administration (FDA) at a zero-tolerance level, which not even the notorious *E.coli* 0157:H7 enjoys.

Escherichia coli

These ubiquitous bacteria, *Escherichia coli* or *E. coli*, are normal human commensals and a most important model organism in research labs. But a number of strains cause various kinds of disease. *E. coli* is the leading cause of urinary tract infections, for instance. It is also behind several forms of gastroenteritis.

The bugs that cause these GI diseases look very much like normal gut flora, but they have acquired additional virulence factors. These bacteria include …

◆ Enterotoxigenic *E. coli* (ETEC), which produces a toxin that leads to watery diarrhea with little or no fever.

◆ Enteropathogenic *E. coli* (EPEC), which causes bloody diarrhea with fever, often chronic and persistent.

◆ Enterohemorrhagic *E. coli* (EHEC), such as strain 0157:H7, which produces large quantities of toxins that damage the kidneys and intestinal lining, especially in children, and produces (as the name suggests) bloody diarrhea.

◆ Enteroinvasive *E. coli* (EIEC), which causes milder bloody diarrhea.

ETEC is said to be the chief cause of traveler's diarrhea. It is most common in developing countries, especially in infants; the source is often contaminated water or food handlers. This infection causes an estimated 600 million cases of diarrhea yearly and 700,000 deaths in young children, almost all in developing countries.

EPEC occurs most often in infants and children in developing countries.

EHEC toxins resemble those of *Shigella dysenteriae*, which we'll get to in a moment. *E. coli* 0157:H7 puts in only occasional appearances, but they tend to be dramatic and get into the news. It is best known for a 1992–93 outbreak in four western states caused by undercooked burgers from one fast-food chain. More than 500 people were reported ill and four died. Cattle appear to be its host organism, and it enters the food supply during processing of beef carcasses.

Other foods also have caused outbreaks, including lettuce, alfalfa sprouts, unpasteurized fruit juice, dry-cured salami, cheese, and raw milk. These are due to cross-contamination, such as rinsing the produce with infected water, cutting the vegetables on a surface where raw beef has rested, or from the items picking up the pathogen from the environment. The disease is sometimes known as hemorrhagic colitis.

Shigella

Some infections are occasionally foodborne, including *Shigella*, *Giardia*, and *Cryptosporidium* (both discussed in Chapter 16), and hepatitis A (discussed in Chapter 14). Disease-producing *Shigella* species not only invade intestinal cells directly, but also produce toxins, called Shiga toxins, that increase their virulence by destroying intestinal cells. Unlike other enteric pathogens, *Shigella* is not transferred from other animals; people are its only natural reservoir.

The disease these bacteria cause is called shigellosis and is most common in children under five. It is not among the most important causes of dysentery (severe diarrhea) in the United States, where many infections are asymptomatic. It is estimated to be 20 times more frequent in the developing world, and causes an estimated 600,000 deaths every year.

Clostridium and Botulism

Botulism, caused by a toxin made by the bacterium *Clostridium botulinum*, is rare but very serious. We have already met the *Clostridium* genus of wicked bugs in our discussion of antibiotic-resistant *C. difficile* in Chapter 12 and the causes of gas gangrene and tetanus in Chapter 15. *C. botulinum* toxin produces neurological disease, causing weakness and paralysis—so much so that patients often need mechanical help with breathing. Like some of its relatives, this bug also infects wounds.

> ### Little Did You Know
>
> The experts say that botulinum toxin is the most powerful naturally produced toxin in the world—up to 100,000 times more potent than sarin gas. And botulinum toxin is also the chief ingredient in Botox, the very popular wrinkle treatment—much diluted, we are happy to report.

The most common form of botulism is infant botulism, which fortunately is rarely fatal. In about 15 percent of cases, the affected infant seems to have acquired the infection by consuming honey. The bug can survive in honey and although it's harmless to most people over the age of three, a baby's stomach does not yet contain enough acid to kill it off, and the baby GI tract has not yet been colonized with enough commensals to compete with *C. botulinum* successfully. The rest of the time, the source of infection is unknown.

The better-known form of botulism is associated only with improperly prepared preserved and canned foods such as sausages and home-canned low-acid vegetables. These provide the perfect conditions for the bug to multiply and churn out its toxin.

This Food Is Rotten

Spoiled or rotted food may look awful, but eating it—if you absolutely had to and could force yourself—would not necessarily make you sick. On the other hand, food that looks and smells fine and tastes good can still have you writhing on the bathroom floor in the middle of the night. The microbes that spoil food, the ones that make milk sour and lunchmeat turn green and slimy, are not the same ones that cause foodborne disease.

Food-spoilage organisms make food deteriorate, but they don't necessarily make it poisonous. Spoilage affects food quality but usually not its safety.

The nature of the food itself governs its ability to spoil. These are called intrinsic factors. A food that's loaded with carbs, like breads and jams, will grow fungus. Foods heavy on protein and fat, like meat, will stink due to anaerobic action called putrefaction. In breaking down the protein, the anaerobes produce amines. These are organic compounds with names like putrescine and cadaverine, which ought to give you some idea what the smell is like.

Here are some other intrinsic factors that promote food spoilage:

♦ Low pH, which encourages yeast and molds

♦ Neutral or high pH, which encourages bacteria

♦ Water, which microbes must have

♦ Food structure (ground meat, for example, spoils faster than a roast because it offers more surface area for microbes to colonize)

Intrinsic factors also can keep food from spoiling. Here are some protective intrinsic factors:

♦ Vegetables and fruits come from nature fully shrink-wrapped in peels and rinds.

♦ Many foods—such as fruits, vegetables, milk, and eggs—contain antimicrobial compounds.

♦ Food "accessories" like herbs, spices, garlic, and hot peppers and the sauces made from them also contain antimicrobial compounds, although herbs and spices can sometimes harbor both disease-causing and spoilage microbes.

♦ Unfermented teas, both black and green, contain compounds that act against bacteria, viruses, and fungi.

The environment around food—what the experts call extrinsic factors—is important for food spoilage as well as food poisoning, especially temperature and humidity. Packaging of food also makes a big difference. Packaging that allows in oxygen promotes surface microbes. Carbon dioxide lowers pH, which inhibits some bacteria and encourages others that are less harmful.

Microbes love the food you eat as much as you do. Molds are the microbes that trigger fruit spoilage. They contain enzymes that can breach protective skins and peels, allowing other microbes entry. Carb-loving bacteria initiate the process in vegetables.

Milk that has not gone through high-temperature pasteurization curdles easily in a four-stage process. First, *Lactococcus lactis* produces acid, which encourages the growth of acid-loving *Lactobacillus.* Then yeasts and molds take over, breaking down the accumulated lactic acid. Protein-loving bacteria can grow when this acidity is reduced, and they make the milk foul smelling and bitter tasting.

Food Preservation

People have been stashing food away for the future for as long as there have been people. That means figuring out how to keep the food safe to eat while stored, which means keeping microbes at bay. Most food preservation is aimed at creating conditions in which microbes cannot survive, or at least making the surroundings so unfriendly that they fail to thrive. Food preservation protects food both from the deterioration of spoilage and the contamination that can lead to illness.

There are many techniques for preserving food. Some are ancient, and some invented only recently. Boiling, for example, has long been an extremely efficient and effective way of destroying microbes in food; it is a type of sterilization with heat, which we discussed in Chapter 11. In the past couple of centuries, heat sterilization in the form of canning has made fairly long-term food storage possible, and pasteurization routinely destroys cow microbes that might make milk drinkers sick.

Another technique is chilling; keeping food cold, or even better, freezing it, is an excellent way to slow down microbial growth. This approach did not have to await the development of electric freezers; the Aleuts and others who live in the far North have been freezing their food for centuries.

Here are some additional ancient methods for preserving food:

♦ Drying or curing of meat, fruit, or cereals because getting rid of water, dehydration, prevents microbial growth.

♦ Adding acid because most microbes can't live in acidic conditions.

♦ Fermentation by, for example, adding lactobacillus to milk, turning it into yogurt that is too acidic for harmful bacteria.

♦ Pickling by immersing food in vinegar, salty brine, ethanol, or vegetable oils and often also boiling them.

♦ Addition of nitrites and/or nitrates has been around for at least a few hundred years. "Salt peter," a common source of nitrites and nitrates, was routinely added to sausages and hams in the curing process.

Here are some relatively new ways of extending the life of foods:

◆ Vacuum packaging (often used for nuts) to get rid of oxygen-containing air because many microbes must have oxygen

◆ Irradiation, treating foods with ionizing radiation to kill microorganisms and cells such as spores, a technique that is controversial because, although food authorities say it is safe, that claim is questioned by some consumers

◆ Addition of new forms of preservatives such as benzoate and citrate that prevent the growth of certain microbes

Plants and Animals Get Diseases, Too

We don't have space to explore animal and plant infectious diseases in the detail they richly deserve. But let's spend a moment at least reflecting on how important the health of other organisms is to our own. At last count, well over half of the 1,400-plus human diseases are zoonoses, and most are found in more than one kind of animal. Zoonoses, you'll recall, are infectious diseases that can be transmitted between people and animals, or can be shared by them.

As we write, experts on flu are deeply concerned with the progress of a bird flu virus that is now infecting humans and are wondering whether it will soon acquire the ability to hop easily from person to person. That is just one of thousands of instances of animal diseases jumping the species barrier to infect our own. And there are many diseases that can't infect us directly, but when they infect the plants and animals we eat, they can have a devastating effect on human life nonetheless.

Plant Pathogens

When plant diseases affect crops, the economic losses can be huge, and more important, people can starve. Most infectious diseases of plants are due to fungi. More than 5,000 fungus species are known to cause plant diseases. These tend to be called mildews, rusts, molds, and root rots. Viral plant pathogens number at least 700. Leaf spots and wilts are bacterial diseases.

The largest group of fungi involved in plant diseases are the Ascomycota, also called the sac fungi, which account for three out of four fungi that scientists have described. This enormous group includes fungi that form lichens, yeasts, mushrooms, the mycorrhizae on plant roots, and *Penicillium*. They live by using potent enzymes to digest organisms, living and dead, and are the world's most important recyclers of material from dead plants.

The strawberries on this plant are completely engulfed by the mold Botrytis cinerea.

(Courtesy of Scott Bauer.)

Here are just a few examples of fungal diseases of major concern:

♦ Soybean rust, caused by *Phakopsora pachyrhizi* and *P. meibomiae,* has reduced yields and increased costs everywhere in the world. It affects more than 90 legume species. The disease has not yet had serious impact in the United States, but that may be because it only appeared domestically in 2004.

♦ *Aspergillus flavus* produces aflatoxin, which contaminates stored nuts and grain. Aflatoxin is the most powerful natural cancer-causing agent known.

♦ *Cryphonectria parasitica* has killed millions of chestnut trees in the United States; the disease is known as chestnut blight. *C. parasitica* entered the United States by accident early in the twentieth century. Less than 40 years later, the American chestnut was essentially extinct.

Animal Pathogens: The Challenges

In the United States, although the number of farms has been decreasing, the number of livestock is increasing all the time, providing new opportunities for contagious disease. In addition, there are completely new sources of food animals. Try finding a creature for sale at the seafood counter that is not a product of aquaculture somewhere in the world. Organically raised animals are now in the meat case in most supermarkets. These changes may well generate exotic new diseases and new disease organisms. They have certainly meant a new interdependence (and new opportunities for transfer of microbes) among the nations of the world.

This is not a new problem, it's intensifying an old problem. Microbes do not respect borders and they are almost never completely eradicated. Take, for example, exotic Newcastle disease, a fatal viral contagion of all birds. A 1971 epidemic in California that affected more than 1,300 poultry flocks had a major impact on the U.S. poultry and egg industries. Nearly 12 million birds were destroyed in an effort to get rid of the disease.

Wild birds are infected with this virus, and so are pet birds, some of them illegally imported. The disease continues to pop up in U.S. diagnostic labs, but mostly has been controlled. A major epidemic occurred in 2002; nearly 20,000 premises were quarantined and well over three million birds destroyed for control. This epidemic, which had affected U.S. trade with other countries, appears to have originated from pet birds in southern California.

Foot-and-mouth disease also is caused by a highly contagious virus that infects cloven-hoofed animals, domestic and wild. This disease has not appeared in the United States since 1929. The United Kingdom was also free of it until 2001, when the virus arrived there in an illegally imported meat product. The result was 6.5 million animals killed for control; economic losses were estimated at 6.3 billion pounds sterling, and British trade was severely affected. The epidemic also spread to the Netherlands.

The Emergence of Emerging Diseases

Emerging diseases usually emerge from animals. Three out of four of emerging human infections in the past few decades have been zoonoses; that is, they have been caused by animal pathogens.

What is an emerging disease? You might at first think: a brand new infectious ailment that no one has ever seen before. And you would be right, but that's not the whole story. A 2003 report from the U.S. National Academy of Sciences declared that an emerging disease "is either a newly recognized, clinically distinct infectious disease, or a known infectious disease whose reported incidence is increasing in a given place or among a specific population."

These emergences are happening not only because new pathogens have evolved, although they certainly have. They are happening also because endemic diseases are infecting increasing numbers of victims, and because microbes are increasingly clever at resisting antibiotics and other drugs. There is also the potential for turning disease organisms into weapons for biowarfare or bioterrorism, a topic we'll take up in Chapter 19.

Some Emerging Diseases

By that definition, the previous chapters are full of emerging diseases, even though some of them, like tuberculosis (TB) and malaria, have been with humanity for hundreds or even thousands of years. TB and malaria make official lists of emerging diseases because they have popped up in new areas or have surfaced in new, drug-resistant forms.

AIDS, on the other hand, is a genuinely new disease, the human immunodeficiency virus having materialized in epidemic form seemingly out of nowhere in the 1980s. It's a remarkable story: in just two decades, this previously unknown disease has become the world's fourth leading cause of death.

Here are some bacterial diseases often considered to be "emerging":

- Anthrax, a possible bioterrorism agent

- Cholera due to the newly evolved *Vibrio cholerae* 0139

- Diphtheria, massive outbreaks due to reduced vaccination

- Hemorrhagic colitis due to newly evolved *Escherichia coli* 0157:H7

- Lyme disease, increasing because of reforestation and a growing deer population

- Plague, a possible bioterrorism agent

- Vancomycin-resistant *Staphylococcus aureus*, which just appeared in 2002

Viral diseases also qualify, among them:

- Dengue because Asian and Latin American epidemics are increasing in size and severity

- Ebola and Marburg hemorrhagic fevers, which we discuss in a moment

- Enterovirus 71 emerged in 1974 and evolves new types quickly

- Hantavirus, increasing because of a growing rodent population thanks to greater rainfall and changes in land use

- Avian influenza or bird flu, which we've mentioned several times

◆ Noroviruses because a recently emerged strain may be more easily transmitted

◆ West Nile, of course

Also on the list is the prion disorder variant Creutzfeldt-Jakob disease, the human version of mad-cow disease. We discussed it in Chapter 4.

A great many complex factors lie behind these emergences, and exploring them all would take several more books. Many of them boil down to human behavior: changing ecosystems; economic development and changes in land use; human population movements and other human behavior (for example, sexual behavior); international travel and commerce; collapse (or nonexistence) of public health systems; dirty water and lack of basic sanitation (especially in the underdeveloped countries); poverty and social inequality; war and famine; and politics, politics, politics.

And then there are the biological factors: the ability of microbes to change and adapt quickly, host susceptibility to infection, failures of some medical technology, climate and weather.

Ebola and Marburg Hemorrhagic Fevers

Just where Ebola and Marburg came from remains a mystery as we write. The viruses are transmitted between people mostly by intimate contact. The original Marburg outbreak was in 1967, when European laboratory workers were laid low by a virulent hemorrhagic fever. They acquired the virus by handling blood and tissues from African green monkeys imported from Uganda. Other outbreaks have also come from monkeys but, like us, nonhuman primates are simply targets for the viruses. Their natural reservoir, animals the viruses infect but do not kill, is thought to be bats.

Ebola hemorrhagic fever surfaced in 1976 in Zaire and Sudan. The symptoms of the two diseases are almost identical.

Marburg and Ebola are, for the moment, the only known members of a new virus family, the filoviruses, RNA viruses that appear to be endemic in central Africa. These viruses are so peculiar looking that for a while it wasn't even clear that they were viruses. Filoviruses are about 80 nm in diameter but often very long—as much as 1,100 nm or more, with bizarre branching forms.

Because they are RNA viruses, filoviruses have a high potential for mutation. This worries researchers who are concerned that they may evolve to become even more scary. It also makes them possible candidates for weaponizing, and that (plus unusual media attention, along with mass-market books and movies) has generated enormous interest in the filoviruses.

Luckily filovirus diseases have been uncommon. These are extraordinarily dangerous viruses that have killed nine out of ten patients in some outbreaks and are subject to the highest level of containment—Level 4—in research labs. As we write, not even 2,000 cases of Ebola have occurred worldwide, but there have been more than 1,200 deaths.

Tiny Tips _____

One strain of the Ebola virus appears not to cause human disease, although it can infect people. It is the Reston strain, which seems to be from Asia, not Africa, and can be deadly to monkeys.

And yet there is antibody evidence that, during an outbreak, many people may be walking around with entirely asymptomatic infections, perhaps due to differences in how their immune systems have responded to filoviruses.

Filovirus outbreaks have been magnified by being spread in hospitals through the use of unsterilized needles and lack of basic hygienic equipment like masks and gowns, always an obstacle in underdeveloped parts of the world. These infections would never have happened if the facilities had access to equipment that Western hospitals take for granted.

A transmission electron micrograph of the weirdly shaped Ebola virus.

(Courtesy of the CDC, Dr. Frederick A. Murphy.)

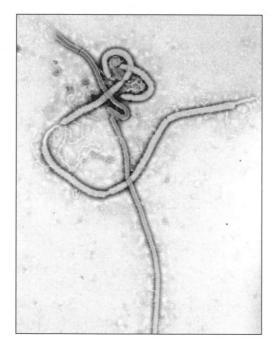

Chronic Diseases as Emerging Diseases

Some authorities think another group of diseases ought to be on the "emerging" list: chronic diseases caused by infectious agents, especially chronic diseases we have traditionally thought of as due to environmental and genetic factors, not microbes.

Microbes and Chronic Disease

We already have discussed several examples, especially those involving cancer. Approaches to cancer research have been revised to take in the idea that these much-feared malignancies are sometimes infectious disease. The bacterium *Helicobacter pylori*, now known to cause most ulcers (an idea that itself revolutionized thinking about this disease), probably causes stomach cancer, too.

And there are several tumor viruses. Most notable is probably the human papilloma virus, responsible for cervical cancer as well as many other cancers, such as head and neck, penile, and anal cancers not to mention both common and venereal warts. The Epstein-Barr virus is behind Burkitt's lymphoma, nasopharyngeal cancer, and Hodgkin's disease. And two of the causes of hepatitis, the hepatitis B and C viruses, also cause liver cancer.

Among the more intriguing ideas are hints that microbes may play a role in behavioral or "mental" disorders, especially if the infection occurred during fetal life. Among the suggestions for which there is some evidence:

- Schizophrenia may result from a viral infection, especially a flu virus, during fetal life.

- Obsessive-compulsive disorder may be linked to infection with *Streptococcus agalactiae*, often regarded as a normal commensal in the human genital tract.

- Congenital mental retardation is linked to cytomegalovirus and Varicella-Zoster virus (cause of chicken pox and shingles).

- Alzheimer's disease may possibly be related to infection with *Chlamydia pneumoniae* sometime during one's lifetime.

A few other fascinating possibilities with some evidence backing them up:

- Multiple sclerosis and Epstein-Barr virus

- Diabetes and enteroviruses

- Amyotrophic lateral sclerosis and prions

Even if such connections are confirmed eventually, it is unlikely that a single microbe is behind all cases of a particular disease. Even for *H. pylori*, where the evidence is very strong, it is clear that ulcers can occasionally have other causes too.

Microbes and Cardiovascular Disease

We spoke briefly about the bacterium *Chlamydia* in Chapter 15. *C. trachomatis* causes trachoma, in developing countries a common eye disease that leads to blindness. In Western countries this bug is more familiar as a cause of STDs and, eventually, pelvic inflammatory disease that can lead to infertility.

C. pneumoniae results in pneumonia, of course, and also colds and bronchitis. But that's not all. There is now evidence that this ubiquitous bug may be a factor in strokes and heart disease.

Most strokes and heart attacks come about because arteries become narrow and constricted, which shuts off blood to the brain and heart. This condition is called atherosclerosis, and it is responsible for half of adult deaths in the developed countries.

High cholesterol levels account for a lot of this clogging, and so does cigarette smoking. But about half the cases of atherosclerosis do not involve these risk factors. Researchers have found that another critical feature of atherosclerosis is inflammation.

As you'll recall from Chapter 10, inflammation is something the immune system does well in response to microbial infection. And, you'll also recall, sometimes it does it too well, and disease occurs because of damage done by the inflammation, rather than directly by the infecting microbe.

Researchers suspect that may be the case in atherosclerosis. They think that clogged arteries can result from a too-fierce immune system response to infection, especially if the infection goes on at a low level for a long time. Six out of ten people show signs of past *Chlamydia* infection, which could help explain why heart attacks and strokes are so common.

This notion is backed by research showing that *C. pneumoniae* infection speeds up the narrowing of arteries in animals. The connection has yet to be firmly demonstrated in people. Research shows that people with heart disease are more likely to possess antibodies against *C. pneumoniae* than people without, and the bug has been found in atherosclerosis lesions.

As it happens, *C. pneumoniae* is not the only microbe researchers think may be involved in cardiovascular disease. Among some other suspects: herpes simplex virus, cytomegalovirus, Coxsackie virus, and the fungus *Histoplasma*.

Preliminary studies also suggest a connection between gum disease, an inflammatory condition, and heart disease. People with gum disease are more likely to have heart disease, as are people who show antibody response to gum disease.

The Least You Need to Know

◆ Food poisoning is gastroenteritis after consuming food or drink that contains particular pathogens or their toxins, often transported in particles of feces or in the food itself. Most food poisoning results from improper food handling, either not cooking it properly or not keeping it at the proper temperature afterwards, and most of it occurs in food from restaurants or institutions.

◆ Spoiled food is unattractive and often tastes and smells terrible, but spoilage is due to different microbes than the ones responsible for food poisoning and may not cause illness.

◆ Plant and animal diseases have a large impact on human health; most human infectious diseases originally came from animals.

◆ Emerging diseases include new diseases as well as existing diseases that are changing their traditional patterns and infecting more people, for example, because pathogens are newly able to resist antibiotics.

◆ Chronic diseases caused by infections may also count as emerging diseases; this list includes many cancers, but also diseases not usually thought to be infectious, such as cardiovascular disease and perhaps even diabetes and some mental illnesses.

Part 5

More Life on Earth, and Under It, Too

This section explores the microbial life in the seas, in our water, and in our soil. We know enough to know that we don't know much about it; it's a vast, invisible universe waiting to be explored. Then we'll move on to biotechnology, a very old way of using microbes to make products that is bursting out in all kinds of new directions. We'll also cover bioweapons: what microbes might be used by bioterrorists and in future wars. Finally, we explore some of the new frontiers in twenty-first century microbiology: probiotics, modulating the immune system to fight disease, microbial forensics, and designer microbes from synthetic biology.

Chapter 18

Microbes on (and in) the Earth

In This Chapter

- ◆ Microbes in the oceans
- ◆ Microbes in fresh water
- ◆ Clean water and how to get it—with the help of microbes
- ◆ Soil—don't call it dirt!
- ◆ Soil microbes and biological crusts
- ◆ Compost and composting

We told you microbes were everywhere. In this chapter, more about microbes in the water (salt and fresh) and in the soil.

The Water of Life

Earth's oceans cover more than 70 percent of its surface. Even though 98 percent of the oceans' biomass, the total weight of all their organic matter, is microbes, much of this vast community remains unknown, even to microbiologists. As one expert has observed, "The forests of the sea are microscopic."

Tiny Tips

Marine microbiology is also called microbial oceanography.

Marine microbes are hard to culture in the lab and therefore have been hard to study, but advances in genetics and other methodologies now provide new ways to explore the depths of the oceans as well as the shallow water, too.

Microbes Go To Sea

The population density of ocean microbes is, forgive the pun, hard to fathom. The surface waters teem with millions of them in every milliliter (ml) of seawater (a little over 0.20 of a measuring teaspoonful). And even in the depths, there are tens of thousands per milliliter.

Almost all photosynthesis is carried out in the top 600 feet. That activity used to be attributed mostly to algae, but now researchers know that the major energy producers are cyanobacteria, which are smaller but far more plentiful.

There are up to 100,000 members of the cyanobacteria genus *Synechococcus* per milliliter. But in the same smidgen of seawater, there are hundreds of thousands of their relative, *Prochlorococcus*, and some of them can photosynthesize even in low light. *Prochlorococcus* is now estimated to account for half of the chlorophyll-based biomass in the ocean.

But the single most successful bug in the ocean appears to be SAR11, soon to be officially named *Pelagibacter ubique*, the well-named ubiquitous bacterium of the open sea. Its ubiquity varies. Near Bermuda, it accounts for half of the bacteria in each milliliter, up to 300,000. And near the U.S. coast, the numbers reach 1 million per ml. Members of this group are believed to constitute about 25 percent of *all* microbial cells.

To achieve such numbers, SAR11 must be doing something right, but no one is sure exactly what it is. One clue emerges from its size, which suggests that it may be the most efficient organism on Earth.

SAR11, which is probably an aerobic heterotroph, is incredibly tiny, a comma-shaped thing 1/100th the size of a typical bacterium. That is very near the size that scientists believe may be the smallest possible for an independently replicating organism. It also has the smallest genome. *P. ubique* is equipped to make all 20 essential amino acids but not much else. It contains no noncoding regions, no transposons, no extrachromosomal genetic elements, no viral genes—no junk DNA of any kind. *P. ubique* appears to be *unique* as well as *ubique*.

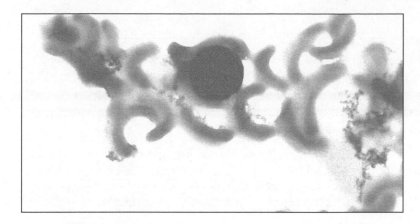

An electron micrograph of SAR11, which appears to be the most successful marine microbe of all. The dark circle in the middle helps make clear the scale of these tiniest of bacteria. The circle is 1/2000th of a millimeter.

(Courtesy of the Oregon State University Laboratory for the Isolation of Novel Species.)

Plankton, Viruses, and Sediment

We met the term plankton when we described the tiny organisms that are the foundation of the ocean food chain in Chapter 3. Plankton consist of many kinds of creatures, but most are microbes. The smallest were just discovered in the 1980s.

There are nanoplankton (between 0.05 mm and 2.0 microns in size), and, the smallest cells, picoplankton (between 2.0 and 0.2 microns across—as little as 1/500th the diameter of a human hair). Picoplankton can double in biomass every day; the reason they don't cover the planet is that protozoa scarf them down just as quickly as they grow.

Viruses—100 million of them per milliliter—are sometimes called femtoplankton. Most are phages, bringing death to bacteria and making their constituent molecules available to others. Each phage attacks a certain kind of bacterium, which keeps any one bacterial group from dominating. Thus phage help maintain the diversity of microbes in the seas.

A major group of oceanic viruses are picorna-like viruses that appear to prey particularly on phytoplankton, including algae that can form toxic blooms. Another group, the cyanophage, shape populations of the cyanobacteria *Prochlorococcus* or *Synechococcus*.

Little Did You Know

What lies beneath the microbes in the oceans? More microbes. Are you surprised? By this time you shouldn't be. The sediments at the bottom of the oceans are teeming with microbes, too—even as much as half a kilometer down. Not much is yet known about the microbes in sediment, but it appears that the bacteria and archaea that dominate this sea subfloor may be quite different from oceangoing microbes.

Fresh Water

Only 3 percent of Earth's water is fresh water, and most of that is locked up in the ice caps and glaciers. Yet fresh water, defined technically as water that contains less than 0.05 percent dissolved salts, is what we need for drinking, and for our crops. The United Nations estimates that well over two million people die every year from drinking contaminated water or from drought.

For obvious reasons, we tend to know much more about pathogens in water, and how to get rid of them, than about the normal nonpathogenic microbial life in rivers, ponds, and even puddles. Microbes in fresh water sometimes have a tougher time than microbes on land or in the seas because the resources are less plentiful.

Pond life is diverse and dynamic. An ordinary pond is stocked not only with many kinds of plants and animals but also with microbes galore. There are photosynthesizing algae and cyanobacteria as well as other photosynthesizing bacteria that do not produce oxygen as a byproduct. One example is the purple bacterium *Rhodospirillum rubrum*. And of course, there are amoebas and other protozoa around to eat the bacteria. Sulfate reducers, which breathe sulfate instead of oxygen, live in pond sediments. And farthest of all from the sun live the methane producers. We discussed some of these methanogens in Chapter 9, but that locale, you'll recall, was the stomachs of cows.

> **Little Did You Know**
>
> Priest Pot is only about 2.5 acres in size, but more than 50 years of research has made it one of the most-studied bodies of water on Earth, especially with respect to protozoan ecology. Scientists have identified nearly 1,300 eukaryote species in this little pond alone.

As you might expect, the complement of particular microbes varies depending on the nature of the body of water. Atlantic white cedar wetlands—acid, low in nutrients and light—attract quite a different microbial constituency than, say, Priest Pot, a small lively pond in England's Lake District.

One revelation of studies like these is that microbial diversity may be much, much greater than previously thought. The ancient protozoan "species" *Neobodo designis* is found all over the world in different marine and freshwater habitats. But recent studies of the DNA of many strains has found so little overlap that it seems likely that there are hundreds, perhaps thousands, of different kinds of *N. designis*.

Sewage, Water, and Pollution

Advanced societies are advanced in part because they provide the people who live there with clean water. Advanced societies also have industries and agriculture and people, and all of these things make water unfit to drink or use for other essentials, like growing food. So by definition advanced societies cannot exist unless they have found ways to clean up their water.

Clean water is also one of humanity's major medical advances. Along with basic sanitation, pathogen-free water has done more to prevent disease than any other technology, including vaccines and antibiotics.

Water Treatment: To You, It's Poo, but Microbes Think It's Yummy

So how do advanced societies achieve clean water? They often call on microbes to do the work for them. Treatment of sewage and wastewater is carried out mostly by our friendly Bugs & Co., in human-devised processes that mimic, adapt, and control the cleaning methods that go on naturally in our rivers, lakes, and other Earthly waters.

There are usually three main steps to water-cleaning technology:

1. Removal of solids by methods such as screening, settling, and adding coagulants that make them clump together and sink.

2. Microbial removal of dissolved organics by biotechnologies such as lagoons, the activated sludge process, and anaerobic digesters.

3. Microbial and chemical removal of inorganic nutrients such as nitrogen and phosphorous, plus removal of viruses.

In conventional sewage treatment, the first step takes place in big man-made ponds or basins, but it also can go on in "constructed wetlands" that employ reeds and other plants, plus their attendant microbes.

In the second step, organics are consumed by microbes, which grow and release carbon dioxide. The point is for the microbes to gather themselves into organized communities called flocs, which then settle out and can be removed, leaving fairly clean water behind. A constant problem, however, is the growth of bacteria that trail yards of filaments, forming big disorganized flocs that refuse to settle out.

These processes all leave behind massive amounts of microbes, living and dead, plus residual organics. This stuff is known as sludge and is subjected to further processing in anaerobic digesters. These huge tanks contain microbes that ferment the sludge, including methanogens to turn it into a dry compact form—not unlike the process we described in the stomachs of cows in Chapter 9. Compacted dried sludge can be disposed of in landfills and by other methods, but since it often contains residual heavy metals and other toxic products, and even pathogens in the form of protozoan cysts, sludge disposal is a source of continuing controversy.

Soil scientist Norm Fausey collects bottom sediments for testing from a constructed wetland built to help clean water draining from a crop field. Microbes in the wetland remove pollutants like nitrites, making the water clean enough to recycle back into the field for irrigation.

(Courtesy of the Agricultural Research Service, Peggy Greb.)

The final step removes material that can't be broken up any further, like some plastics, toxic heavy metals and nutrients like nitrogen and phosphorous that promote eutrophication. This stage employs a combination of chemicals, filters, and microbes. Water intended for drinking gets additional treatment—usually the addition of chlorine, among other chemicals, which kills most pathogenic microbes.

The activated sludge process is the most important aerobic system for treating wastewater. Activated sludge is mostly billions of bacteria from hundreds of different species. Microbes not only remove the sewage, they also get rid of the nitrogen and phosphorous.

The activated sludge process has been around for a century or more, but only recently have microbiologists learned much about how it works. They have discovered, for example, that the filamentous bacteria that cause problems with floc are much more diverse than previously thought—a finding that may eventually lead to better ways of controlling them.

Toward Clean Water

Municipal and other water treatment systems get rid of most of the potential pollutants from sewage, industrial waste, and agricultural runoff. But a lot escapes capture too: most agricultural runoff, animal wastes, wastes from inadequate treatment systems, and inadequately controlled industrial waste. Pollutants can end up in rivers and groundwater, drain into aquifers and the sea, and be carried to distant communities.

Microbes are pressed into service to help with these problems too. *Escherichia coli* and some of its relatives (called coliforms), and *Enterococcus* are the standard for "indicator" microbes, serving the function of the canary in the coalmine.

Because they normally make their homes in the human gastrointestinal tract, their presence is a warning that the water may contain human feces. If these bugs cannot be detected in a certain volume of water (usually 100 ml), the water is considered potable or safe to drink. In Chapter 20, we talk about experiments with microbial cleanup of water and pollution that may end up in widespread use some day.

The Community of Soil

Soil is another one of those microbial communities like the ones we discussed in Chapter 9. It's a community second only to water in importance for life on Earth. Microbes in the soil clean the water and air, and help plants grow. And plants provide food for all, plus oxygen, while they store carbon dioxide and nitrogen.

Soil: Don't Call It Dirt!

Soil is made of minuscule bits of rock that have weathered off big rocks and combined with organic material and water. Depending on the proportion of these basic ingredients, soil composition varies enormously from place to place.

Soil contains a mind-boggling number and variety of organisms, which live in and between soil particles. Many (such as nematodes and arthropods) are not microbes, so we'll ignore them except to say that they are crucial to soil too.

> **Little Did You Know**
>
> Soil microbes have different and overlapping functions. Photosynthesizing bacteria and algae capture energy. Bacteria and fungi break down soil components, boost plant growth, and, sometimes, cause disease. Protozoa consume bacteria and contribute their components.

But soil contains all the microbes: bacteria, archaea, viruses, protozoa, algae, and fungi. Scientists have estimated that just a teaspoon of soil contains between 100 million and a billion bacteria and several thousand (sometimes a million) protozoa. That means that each acre of soil is home to a ton of bacteria—literally, a ton.

An average soil is about one-half undigested organic matter (which is available for consumption by organisms) and one-half humus (which is mostly digested organic matter). Numerous as they are in soil, organisms account for less than 5 percent of soil organic matter, and fresh plants less than 10 percent. Air and water, necessary for growth, reside in the spaces between soil particles, and water clings to the particles themselves.

Meet the Soil Microbes

Most bacteria consume the simpler compounds in soil, such as sugars from plant roots. This makes the nutrients available to other organisms that eat the bacteria. Some bacteria also can decompose pollutants and pesticides. Others, as you'll recall from Chapter 9, form symbiotic relationships with plants that work for them both; these are the rhizobia. Still others, the lithotrophs or chemoautotrophs, make their own energy from inorganic (non-carbon) compounds rather than sunlight, a process that assists nitrogen cycling and also reduces pollution.

Bacteria help water move through the soil by binding soil particles together; this also helps soil hold water longer. Nonpathogenic bacteria also compete with bacterial pathogens in the soil, which reduces disease. Some scientists believe that plants in a particular location (and plant pathogens, too) can be managed by managing the attending soil bacteria.

Little Did You Know

Most of our antibiotics, you may recall, originated in soil bacteria, which have warred with these chemical weapons for billions of years. Researchers recently tested nearly 500 different soil microbes for resistance to antibiotics. The very bad news: every single one of them, all from the genus *Streptomyces*, resisted at least six antibiotics, and some bugs could laugh off as many as 20. Several of these resistance mechanisms had never been seen before.

Fungi consume the more complex compounds in soil, such as plant fibers, and subject humus itself to further decomposition. Their hyphae, which can be yards long, push through the soil and bind particles to each other, helping to hold on to water. And just as there are symbiotic bacteria that help plants grow, you'll recall that there are also symbiotic fungi that partner with plants, the mycorrhizae. Of course, there are also fungal pathogens, such as *Verticillium*, a root pathogen. Fungi can also control diseases by trapping nematodes that cause them.

Like plants and cyanobacteria, soil algae photosynthesize to make their own food, making atmospheric nitrogen and carbon dioxide. They also make phosphorous available to plants. The photosynthesizers comprise about a quarter of the biomass on agricultural land and are the dominant microbes in some watery agriculture like rice paddies. Soil algae also can remove pollutants like heavy metals and excess nutrients. This process is called phycoremediation, or algal phytoremediation.

Protozoa eat a lot of different things, but they particularly like bacteria, which is why you'll find them clustered around plant roots, hanging out where the bacteria hang out. A ciliate can scarf down as many as 10,000 bacteria in a day. As they consume bacteria, protozoa free nitrogen for use by the other organisms. One group of protozoa eat fungal root pathogens.

The world's oceans are now known to be full of viruses. But very little is known about the normal complement of viruses in soil, except for some viruses that cause diseases, especially plant diseases.

One small survey attempting to learn something about indigenous virus populations in soil has found that viruses are most abundant in wetland soils, and more abundant in forest soils than in agricultural soils. In short, the more water, the more viruses.

As is the case in the ocean, most soil viruses were phages. There were also hints that the virus numbers correlated with bacteria numbers in various soils, which might help explain why phages, which prey on bacteria, were relatively common in this limited survey.

Like all archaea research, studies of archaea in soil are at an early stage. But these investigations so far have helped to reveal that these organisms live everywhere, not just hot springs and deep-sea vents. One surprise has been that members of the *Crenarchaeota*, a major group of archaea known for liking it hot, also are present in the oceans and soil at ordinary temperatures.

Archaea have been hard to study because they are hard to culture in the lab. But studies of their genes have shown, for example, that at least two kinds of *Crenarchaeota* colonize the roots of tomato plants growing in Wisconsin. Other studies suggest that the archaeal population dwelling among plant roots varies depending on the soil and the roots. Nobody knows exactly what these archaea are doing in the soil, but the assumption is that they wouldn't be there if they weren't doing something.

Biological Soil Crusts

In places where it is dry and there isn't much plant life, like deserts, the part of soil is played by biological soil crusts—actually a whole ecosystem, a community dominated by lichens, cyanobacteria, algae, mosses, and sometimes fungi and bacteria, bound together with soil particles. These living soil crusts are not to be confused with other kinds of crusts, such as a salt crust. The biological crusts are sometimes called cryptobiotic or microbiotic soil.

Wee Warnings

Crusts can tolerate tough conditions but don't like to be trod upon by cattle, people, or anything else.

Biological soil crusts are full of mini-hills called pinnacles that can be as much as 10 cm high and are formed by filaments from cyanobacteria and green algae. This rough surface helps hold water. The crusts extend only 1-4 mm underground, but still provide soil stability, erosion prevention, nitrogen fixation, nutrients, and a friendly place for seeds to germinate.

Where the Soil Microbes Live

Most soil organisms (microbes, plants, and all) live in the top few inches of soil, which is called—surprise!—topsoil. Beneath is subsoil, which is dense and sparse in organic material.

Bacteria tend to cluster around plant roots and gather the abundant food there. Protozoa do too, because, as we have said, they eat the bacteria. Fungi are happy in leaf litter, which gives them a lot of complex material for decomposition. They like humus too, because only fungi possess the right kind of enzymes for decomposing the already-decomposed humus even more.

Each bit of land is a community unto itself, with a unique combination of organics, minerals, and organisms. Prairies and grasslands tend to be dominated by bacteria, and so do agricultural soils, although exceptionally productive agricultural land also houses a high proportion of fungi. Forest soils are dominated by fungi, with more fungi than bacteria in a deciduous forest and many more among the conifers.

The mix of protozoa varies by ecosystem: a teaspoon of agricultural soil possesses thousands of amoebas and flagellates and only hundreds of ciliates. In a teaspoon of forest soil, you'd find hundreds of thousands of amoebae and fewer flagellates. Land management also governs these mixtures. Land that is not tilled contains more fungi in proportion to bacteria than heavily tilled agricultural land.

Compost: Fine Dining for Microbes

Compost is really just another word for humus: decomposed organic matter that is good for the soil and for plant growth. Compost is brought to you by microbes, the essential recyclers, which do the decomposing.

One reason compost helps plants grow is because it helps mycorrhizae grow. Plant pathologist Robert Linderman and technician Anne Davis hold mycorrhizae-inoculated marigolds. Linderman's larger marigold received compost, while Davis's smaller one did not.

(Courtesy of the Agricultural Research Service, Peggy Greb.)

These microbes form a community and transmute dead plants, dead insects, the remains of last night's dinner, and other garden or forest organics into forms that can be reused by living organisms. Most compost microbes are bacteria, but fungi and protozoa also contribute as do larger creatures like worms.

Composting is the ultimate in a natural recycling process, but people have devised ways of controlling it. Some have turned compost control into a science. Serious composters seek to provide their microbe zoo with optimum conditions for making compost as quickly as possible.

That means certain ratios of high-carbon ingredients (like dead leaves) to high-nitrogen organics (like freshly cut grass), in a proportion of about 25 to 1. Serious composters have serious arguments about these proportions.

That also means providing the decomposing compost with oxygen by turning it periodically; this discourages anaerobes, which produce bad smells when they do their decomposing. And providing it with water, keeping it damp but not sopping. And it may mean encouraging the kinds of bacteria that like heat and will produce it. Heat kills pathogens and weed seeds and speeds up decomposition.

Other composters are serious too—serious about spending time on activities other than composting. They just toss their plant material and garbage into an out-of-the-way corner of the yard and go do something else. Their compost will be made a bit more slowly, but the microbes will turn up just the same and do the same fine job of recycling.

The Least You Need to Know

♦ Marine microbes are hard to culture in the lab and therefore have been hard to study, but advances in genetics and other methodologies now provide new ways to explore microbes in the ocean and other environments.

♦ Microbial life in fresh water is also diverse and dynamic, but we know much more about pathogens in water than we do about harmless microbes.

♦ Basic sanitation and pathogen-free water have done more to prevent disease than any other technology, including vaccines and antibiotics.

♦ Treatment of sewage and wastewater is carried out mostly by microbes, especially bacteria, in human-devised processes that mimic, adapt, and control the cleaning methods that go on naturally in Earth's waters.

♦ Soil is a microbial community second only to water in its importance for life; microbes in the soil clean the water, clean the air, and help plants grow.

♦ In the soil, photosynthesizing bacteria and algae capture energy, bacteria and fungi break down soil components, boost plant growth, and sometimes cause disease, and protozoa consume bacteria and contribute their components.

Chapter 19

Microbes Go to Work— and to War

In This Chapter

- Making microbes work: biotechnology

- Fermentation

- Agricultural, medical, and industrial biotechnology

- Biotech controversies

- Biological warfare and bioterrorism

- Bioweapons, past and future

We can hardly convey in this brief survey the magnitude of biotechnology. It's an old human invention, but also showcases the newest of the new. In this chapter, we give you a bit of both.

Biotechnology: Making Microbes Work

Biotechnology can be defined pretty simply: it just means using living organisms to make products. The term may be simple, but it covers an enormous range of activities. Here are some examples of biotechnology:

◆ Making beer (probably the first human biotechnology)

◆ Genetically manipulating a cow to get her to produce milk full of medicines that can be extracted to cure human disease

◆ Genetically engineering a plant to make it pest-resistant

◆ Producing specialized enzymes for use in products, from paper-making to laundry detergent

Several prospects for biotechnology we describe are just that at this point: prospects. They are being researched, but most are not yet available. Still, others are already in the marketplace. This is especially true of agricultural and to some extent medical biotechnology. Genetically modified (GM) plants are grown in many places around the world. Vaccines and drugs that are products of biotechnology are preventing and curing human and animal diseases.

Tiny Tips

Not all biotechnology is based on microbes, but a lot of it is. Yeast makes beer out of grain. Researchers use viruses to ferry new genes into cells when they are trying to create genetically engineered animals. Microbes generate hundreds of different enzymes useful in industry.

GM products are also controversial, a topic we discuss more later. This is one of the reasons they have been slow to appear. Companies are trying to anticipate and avoid controversy, so they are cautious and highly selective about which products they will attempt to market.

Technical realities have also gotten in the way of swift progress. It is far more difficult to change the composition of foods to make them more nutritious or less allergenic than it is to insert a single gene for herbicide resistance.

Fermentation: Your Basic Biotech

As examples like beer-making show, some forms of biotechnology rely on fermentation. *Fermentation* is a natural process in which a cell (often a microbial cell) gets the energy it needs by breaking some organic material down into its component parts without using oxygen. Humans can harvest those component parts—including some that the microbes discard as waste—and turn them into things that people use.

The process begins with glycolysis, which we discussed briefly in Chapter 9. Glycolysis is the splitting of a sugar molecule to yield energy. For example, splitting a glucose molecule, which contains six carbon atoms, into two molecules of pyruvic acid, which contain three carbons apiece, generates a bit of waste. It also generates a small amount of adenosine triphosphate (ATP), a molecule that stores and transports energy.

The following are notable facts about glycolysis:

♦ It does not use oxygen, and yet it is the first step in several kinds of chemical reactions that generate energy, both anaerobic and aerobic.

♦ It goes on in the cell's cytoplasm, not in any specialized organelle.

♦ It is the only metabolic pathway found in all major groups of organisms.

These facts suggest that glycolysis may be life's most ancient way of producing ATP, dating back to before there was oxygen in Earth's atmosphere.

The purpose of fermentation—at least from the human point of view—is conversion: conversion of grain to beer, fruit juice to wine, or carbohydrates to carbon dioxide to make bread airy and tender. There are several kinds of fermentation processes, which are named for the main organic end product each produces. Here we describe three, one of which does employ oxygen:

♦ Lactic acid fermentation (carried out by fungi and some bacteria, such as lactobacillus)

♦ Alcohol fermentation (carried out by yeast and some kinds of bacteria)

♦ Acetic acid, plain old vinegar (produced from a dilute alcohol solution, with the help of oxygen, by *Acetobacter* bacteria)

When *Lactobacillus acidophilus* turns milk into yogurt, it converts pyruvic acid into lactic acid, which gives yogurt its tangy taste.

The results of alcohol fermentation are bread, beer, and wine—humanity's oldest biotechnology products, invented many thousands of years ago. All are based on the waste products left behind as yeast or bacteria generate ATP from sugars: carbon dioxide and ethanol. Bread rises due to carbon dioxide's gassiness and smells good while baking as it burns off ethanol. Ethanol is the only drinkable alcohol, and carbon dioxide makes beer (and some wine) fizzy.

Fermentation can sometimes spoil food—for example when *Acetobacter* turned that bottle of wine you opened just last week (or was it the week before?) into something undrinkable. But fermentation has contributed far more to human cuisine than it has ruined.

The number of foods produced and preserved by fermentation is nearly uncountable. In addition to basics such as bread and beer, here are a few other examples:

♦ Numerous dairy products (yogurt, cultured sour cream, buttermilk, acidophilus milk, kefir, and several other fermented milks)

Tiny Tips _____

Most likely you eat or drink fermented foods on a regular basis. Fermented foods include many popular items, such as bread, yogurt, lunchmeat, and pickles. (School lunches would surely be different without fermented foods.) Of course, let's not forget beer and wine, too!

♦ Cheeses (from soft unripened to hard ripened, some 2,000 of them invented in the past 8,000 years)

♦ Cured meats (hams, sausages, cold cuts)

♦ Fish (fish sauce, katsuobushi)

♦ Cabbage family members (sauerkraut, kimchi)

♦ Soybeans (soy sauce, tempeh, miso)

♦ Other vegetables and fruits (green olives, pickled cucumbers, and pickled peaches)

Fermentation can be carried out on almost any scale—from a few milliliters in a test tube to a giant fermenter (also known as a bioreactor) that holds thousands of liters.

Large-scale bioreactors can make everything from medical products such as human insulin produced by genetically engineered organisms to antibiotics made by *Bacillus subtilis*, *Streptomyces*, and *Penicillium*. In addition to microbes, these processes also use plant and animal cells for making some kinds of products.

Little Did You Know

The first scientist to describe details of fermentation was Louis Pasteur, who began studying it in the 1850s. Among other insights, he concluded that oxygen inhibited fermentation, and in the process he discovered anaerobic microbes.

"Green" Biotech: Agriculture

Humanity has been breeding plants and animals for specific purposes for at least 10,000 years, probably longer. So it might be said that genetic manipulation of life-forms for agricultural purposes is nothing new in our history. Almost all of today's agriculture, including plants and animals, involves organisms that have been crossbred or hybridized (or, as we shall see, products of agricultural biotechnology—agbiotech, for short).

Agbiotech is something new. It is not concerned, as our ancestors were, with picking two members of a species with desirable traits, breeding them, and hoping for even better offspring—or with saving seeds from this year's best crop plants for growing next year.

In most cases, agricultural biotechnology seeks to transfer a single new genetic trait into an organism (or, recently, stacking a couple of traits, such as herbicide tolerance and disease resistance, in the same GM organism). Sometimes GM involves insertion of a trait from an entirely different kind of organism—for example, fighting the European corn borer by giving corn a gene from the bacterium *Bacillus thuringiensis*, which makes a protein toxic to the insect.

Thus, agbiotech is much more specific and selective than the agricultural genetics practiced by past generations of farmers.

The range of possibilities for GM is huge. The following are a few examples that resulted from genetic engineering:

- Plants may be given genes that help them resist herbicides, the idea being to spray a field and kill the weeds without harming the crop plant.

- Plants can be genetically engineered to resist diseases and insect pests.

- Conferring more nutritional value on food plants or genes to resist spoilage.

- Reducing allergens in foods by altering their molecules.

- Helping crops survive in arid climates and in poor soil.

- Producing crops that require less labor to grow and harvest.

The earth is already home to more biotech crops than most people realize. In the United States, the majority of cotton and soybeans are modified to resist herbicides, and about half the corn (maize) is engineered to resist disease. In the rest of the world, GM crops are grown by 8.5 million farmers in more than 20 nations, about half of which are developing countries. The most popular crop is GM soybeans, followed by maize, cotton, and canola. The new genes they contain confer herbicide tolerance and insect resistance.

Plant and animal biotech includes activities besides genetic manipulation of crop plants. For example:

- Forest biotech, with the aim of creating disease- and insect-resistant trees that grow faster and spend more of their energy making wood instead of flowers and seeds

- Livestock biotech for increasing the health and productivity (and leanness) of farm-grown food animals, fish, and poultry

- Biotech applied to companion animals, improving the health and well-being of cats and dogs

- Cloning techniques for preserving endangered species, both plants and animals

"Red" Biotech: Medicine

There is a lot of similarity in agbiotech and medical biotech. For example, much of the work that could be called agbiotech could also be called medical: breeding disease-free plants and animals and creating vaccines to keep them that way, for example, are medical applications. Here are some others:

- Creating animals—often pigs—to donate organs or blood to people, a field now with its own name: xenotransplantation

- Pharming, the genetic engineering of mammals—cows, goats, sheep, and even rabbits—to produce molecules in their milk that can be extracted and used for human therapy

- Plant pharming: engineering plants to produce extractable drugs, or edible plants that contain vaccines and other medicines to be consumed by mouth

- Biotech cows to make designer milks with enhanced nutritional qualities, or fewer allergens, or less lactose for the millions of people who can't digest this milk sugar

- DNA and antibody-based tests that permit quicker, more accurate diagnoses of animal and human diseases

- With the help of genome studies of both pathogens and people, improved prospects for specialized vaccines and precisely targeted therapies

- Making new medicines, such as the drugs Ritixin and Avastin for cancer therapy

"White" Biotech: Industry

Industrial biotechnology relies most heavily on genetically engineered microbes to produce products and improve industrial processes, often in fermenters and bioreactors. People have been talking about the promise of industrial biotechnology for decades, and also lamenting that it has moved more slowly than agricultural and medical applications.

In the early days, attention was on pollution: using microbes to clean up oil spills and other environmental toxins. Bioremediation continues to be the fastest-growing segment of the hazardous waste industry.

The biotechnological approach to industrial processes is being taken more seriously than ever before—especially in the chemical and energy industries. These trends are being driven by concerns about the environment and climate change, as well as by rapidly increasing energy costs.

> **Little Did You Know**
>
> Researcher attention has recently turned to biofilms as a tool for bioremediation. One example: creating biofilms that block the flow of groundwater through contaminated sites. Another example: using biofilms that contain metal-scavenging bacteria to clean up streams around mines.

Biotechnology products contribute more than 5 percent of the chemical industry's global sales and are expected to pass 10 percent by 2010. So biotechnology produces antibiotics and it controls biofilms. It makes possible biodegradable plastics that could reduce waste up to 80 percent while reducing the petroleum-based plastics industry oil consumption by well over half.

In particular microbes produce enzymes useful in an enormous number of ways. Enzymes (which, you'll recall, speed up chemical reactions) can replace many chemicals traditional in industrial production.

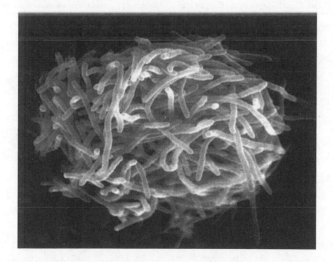

A scanning electron micrograph of the bacterium Acidothermus cellulolyticus, *isolated from a hot spring in Yellowstone National Park. This strain contains several enzymes that help break down biomass into sugars for subsequent fermentation into biofuels and other chemicals. NREL has adapted strict anaerobic microorganisms to degrade a variety of organic wastes to useful by-products, including methane.*

(Courtesy of the NREL, Mark Finkelstein.)

Genetically enhanced enzymes are being used to turn cellulose—essentially plant waste such as corn stalks—into fuel ethanol. In the brewing industry, they can reduce the product's time to maturity. They can take the place of previously used emulsifiers and oxidants in the baking industry. They can replace toxic chlorine for bleaching in the pulp and paper industry. They make ascorbic acid (vitamin C) out of glucose.

Microbial enzymes also make many consumer products, or make them more desirable. Enzymes in laundry detergent get clothes cleaner in cold water, they "stonewash" jeans without damaging the denim, and they churn out high-fructose corn syrup for soft drinks.

The Biotech Controversies

Many people are disturbed at the idea of human beings manipulating the genes of other organisms. Some of their objections are religious—dismay that humans should be "playing God."

Others are worried about safety. Suppose the engineered organism gets free and spreads its genes—giving weeds, for instance, the ability to resist herbicides? How can we be sure that pharmed animals, with the ability to make potent drugs in their cells, stay out of the food supply? Will genetically engineered foods be safe to eat?

Little Did You Know

Another aspect of agbiotech that causes concern is the potential for novel food allergens. For example, at one point GM soybeans contained a gene from Brazil nuts. This protein is potentially allergenic to people with nut allergies. But who would expect that a soybean could trigger a nut allergy? Fortunately, the problem was identified before it reached the mass market.

Still others are concerned about the economics of genetic manipulation. Genetically engineered crop plants usually don't produce fertile seeds. This means farmers can no longer save some seeds from a crop for planting next year; they must buy new seeds every year—a big financial burden for subsistence farmers in poor countries.

To date most of these controversies have swirled around agbiotech. This has even resulted in bans on genetically modified products in some parts of the world. In others—the United States, for example—there seems not to be much concern about genetic engineering, at least with respect to agricultural products.

Medical biotech has generated some disputes in the United States (and elsewhere). The major one has concerned the disappointing record of gene therapy. As we write, it has not yet reported any notable successes, and in a few cases patients have died as a result of attempts at therapy. Genetically engineered drugs and vaccines, however, have met with little or no opposition; people mostly accept them and don't seem to worry about their safety as they have about some GM foods.

The least amount of furor has attended industrial biotech. That may be in part because industry has been slow to adopt GM techniques. Safety concerns may also be less because GM organisms are often confined in giant bioreactors, rather than planted in farm fields.

A Different Kind of Warfare: Bioterror and Biowar

Biological warfare and bioterrorism are different in scale, and one is carried out by nations while the other is the potential weapon of small groups.

The U.S. National Institutes of Health says microbes that make effective bioweapons are likely to have many of the following characteristics:

- They cause a lot of death and disease.

- They can be transmitted from one person to another, sometimes with the help of a vector.

- They require only a low dose to infect, with the microbes causing large out-breaks of disease.

- They can contaminate food and water supplies.

- There is no specific diagnostic test or effective treatment.

- There is no safe and effective vaccine.

- The microbe makes the public and health-care workers anxious and fearful.

- The microbe can be weaponized—that is, turned into a weapon.

These characteristics apply to pathogens or toxins that can contaminate food and water supplies and certain zoonotic agents that can spread infection to humans from domestic animals.

Several emerging and reemerging pathogens, such as West Nile virus, influenza viruses, and drug-resistant *Streptococcus* and *Staphylococcus*, also are potential agents of bio-terrorism. In addition, organisms that don't fit these characteristics in their native state may be engineered by relatively simple genetic manipulations to become a significant threat.

The U.S. National Academy of Sciences observed the following:

The knowledge needed for developing biological weapons is accessible to individuals through the open literature and the Internet; the technology is readily available and affordable; and, perhaps most alarming, as the field of molecular genetics advances, an increased capability exists to bioengineer vaccine- or antimicrobial-resistant strains of biological agents. Currently, many terrorist-sponsoring nations or states are suspected of having active bioweapons programs in place.

Bioweapons, Past and Future

Bioweapons are an ancient warfare strategy, dating back thousands of years. Long ago combatants poisoned the enemy's wells with toxic fungi and diseased animal carcasses and flung the corpses of plague victims over the walls of besieged cities. More recently the tools of research have been employed in attempts to make pathogens more deadly or easier to distribute.

Little Did You Know
Tularemia was turned into an aerosol weapon in both the United States and the former Soviet Union; in the latter, it was also engineered to resist vaccines, according to the National Academy of Sciences. The Soviet Union weaponized at least 30 biological agents, including some others that were resistant to drugs and vaccines.

Biological weapons were formally outlawed in 1972, in an international treaty signed by more than 100 nations. But it failed to stop research on bioweapons, which was carried out by the United States and the former Soviet Union, and other nations, during the Cold War and after.

Both the United States and the former Soviet Union weaponized pneumonic plague. They experimented with methods of delivering the bacillus, including making it deliverable by aerosol—not an easy technical accomplishment, because it is normally transmitted by a vector. They also worked on making it more virulent and resistant to antibiotics, and combining it with other diseases such as diphtheria.

Smallpox and anthrax have been called the poster children of bioterrorism. Both are potentially lethal; both are stable in aerosolized form; and both can be produced on a large scale. In addition, they have no smell or taste and symptoms take some time to develop, giving the terrorists time to get away. Vaccines and treatments are not widely available.

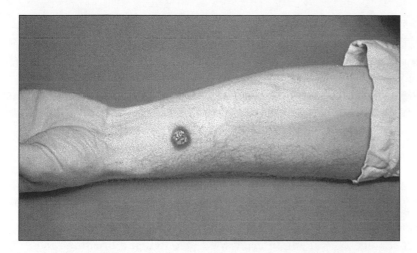

Cutaneous anthrax leaves lesions on the skin, such as this one on a man's forearm.

(Courtesy of the CDC.)

These all contribute to the exceptional psychological impact that is likely to attend any use of these two diseases as weapons. Many people now alive still remember the devastation wrought by smallpox before it was banished to the health authorities' freezers. And we all remember the panic that ensued after anthrax-containing letters began arriving on the desks of journalists and members of Congress soon after terrorists flew planes into the World Trade Center and the Pentagon in September 2001. The anthrax letters killed five people and made six others very ill—and we still don't know who sent them.

What Microbes Make Good Weapons?

In a sense, bioweapons are a variant on the problem of emerging diseases, except that in this case emergence is due to deliberate human acts. A great many potential scenarios are possible.

Pathogens that have been aerosolized—that is, can be distributed through the air—could potentially affect a great many people if weather conditions are just right. Many experts are worried about food-borne pathogens, which can make people sick in low doses and are easy to distribute. Others think that policy makers have paid insufficient attention to the potential for attacks on crop plants and food animals.

The experts have thought about how to describe the degree of threat that various microbes might pose and have divided them into three degrees of risk. The National Academy of Sciences has published a list.

Category A on that list is the highest priority for concern because the organisms "can easily be disseminated or transmitted from person to person, result in high mortality rates, could cause public panic and social disruption, and require special action for public health preparedness." These diseases include the following:

- Anthrax (*Bacillus anthracis*)

- Botulism (*Clostridium botulinum* toxin)

- Plague (*Yersinia pestis*)

- Smallpox (*variola major*)

- Tularemia (*Francisella tularensis*), which is mostly a disease of rabbits and squirrels, but is highly infectious and has a 30 percent mortality rate in humans; it is spread by insects, direct contact, consuming infected food or water, and breathing in the bacteria

- Viral hemorrhagic fevers (filoviruses such as Ebola and Marburg, and arenaviruses such as Lassa fever)

Category B diseases are also cause for concern because they "are moderately easy to disseminate, result in moderate morbidity rates and low mortality rates." They also require specific enhancements of the public health infrastructure's diagnostic capacity—plus enhanced disease surveillance. The items involving microbes include these:

- Brucellosis (*Brucella*)

- Epsilon toxin of *Clostridium perfringens*

- Food safety threats (*Salmonella*, *Escherichia coli* 0157:H7, *Shigella*)

- Glanders (*Burkholderia*), mostly diseases of horses that can infect people

- Psittacosis (*Chlamydia psittaci*), primarily a respiratory infection of birds but they can transmit it to humans and other animals

- Q fever (*Coxiella burnetii*), a highly infectious organism that mostly causes disease in animals, but it can be inhaled by humans, and the chronic form of disease is frequently fatal

- Staphylococcal enterotoxin B, one of those superantigens that can cause toxic shock

- Typhus fever (*Rickettsia prowazekii*)

◆ Viral encephalitis due to alphaviruses that cause Venezuelan equine encephalitis, eastern equine encephalitis, and western equine encephalitis—not just horse diseases, but human diseases as well

◆ Threats to water safety (notably *Vibrio cholerae* and *Cryptosporidium*)

Category C comprises emerging pathogens that could be engineered for wide dissemination because they are available, easy to produce, and have the potential to cause much disease and death. They include hantavirus and influenza viruses, such as H5N1.

Experts say the emphasis on lists of this sort may be misplaced. Policy makers might be misled by history into preparing to defend against the bioweapons of the last war. Perhaps they should not be basing their approaches to today's biothreats on the Cold War's state-sponsored weapons research that relied on industrial processes and official military strategies.

The idea that only some microbes are potential bioweapons does not take account of the newer techniques of molecular biology and genetic manipulation. These methods, they say, may expand greatly the list of organisms that can be weaponized and produced on a large enough scale to spread death and disease.

On the other hand, these same scientific tools give researchers unprecedented power to create new antimicrobial drugs and vaccines and new diagnostics. These can not only protect populations against what bioterrorists may do, but against the afflictions invented by the most successful bioterrorist of all—nature herself.

The Least You Need to Know

◆ Biotechnology is defined as using living organisms to make products; products range from beer to genetically engineered cows.

◆ Fermentation is a basic biotechnological process in which microbes break down organic material to harvest energy in the form of ATP; the process is called glycolysis.

◆ Fermentation can sometimes spoil food, but mostly it is used to produce food (such as yogurt) or to preserve it (for example, cured meats, sauerkraut, and pickles). Fermentation can also produce medical products such as antibiotics.

◆ Agricultural biotechnology, also known as green biotech and agbiotech, is an ancient human practice that began with deliberate breeding of plants and animals more than 10,000 years ago.

◆ Biotechnology is controversial for several reasons. Much of the concern has been about its safety—whether engineered organisms would escape confinement and cause damage or transfer their new genes to other organisms, or whether GM food is safe to eat.

◆ Both biological warfare and bioterrorism involve intentionally using microbes or their toxins to cause death and disease—and spread panic and social chaos—to vanquish an enemy or achieve ideological goals.

Chapter 20

Microbiology in the Twenty-First Century

In This Chapter

- New technologies, new microbes
- Probiotics: edible germs
- Microbial forensics
- Passive and active immunities
- Virtual microbes and synthetic life
- Microbiology as a career

In previous chapters, we have mostly described what is already known about microbes, and how they are being used. In passing, we have also sometimes pondered what the future might hold for this field. For example, we discussed insect genetic engineering to prevent arbovirus disease, the development of specialized bacteria to clean the water coming from mining sites, and how scientists are trying to find new antibiotics. In Chapter 19, we discussed the changes in industrial biotechnology. The experts predict that trend will continue and then some, moving the chemical industry more rapidly away from its traditional manufacture into greener pastures with products that are environmentally friendly.

In this concluding chapter, we peer into the future even more, exploring other places microbiology might take us in the next few years and decades.

The Future of Microbiology

Microbiology is more important than it ever was before. Those of us who live in the developed world have had much of the burden of traditional infectious diseases lifted from us. The rest of the world, as we all know, is not so fortunate.

And yet even in the developed world, the microbes keep coming. Emerging diseases. Biowar and bioterror. Antibiotic resistance. Nosocomial infections. Food contamination. Disease-laden waters.

As its store of knowledge builds up, microbiology itself is changing. The focus is shifting toward the study of microbes in groups, because we realized that microbes do not act as individuals in the real world. The subject matter—even for pathogens—is becoming microbiological ecosystems, the importance of communities of microbes and the ways they work with (and compete with) each other. Infectious disease is itself an ecological phenomenon.

New Technologies, New Microbes

There's a vast universe of microbes out there, and comparatively few have been cultured for individual lab study. New technologies—proteomics, nanotechnology, biocomputation—make it possible to assess entire microbial habitats for the first time. As mentioned in earlier chapters, genomics is crucial. Among other virtues, it can pinpoint the many previously unknown microbes around us, and can also help researchers find new treatments and preventives by identifying pathogen vulnerabilities. New imaging techniques permit far more detailed analysis of microbial cell structure—and the structure of microbial ecosystems.

Tiny Tips

We can expect research eventually to identify what factors trigger new or more virulent disease organisms and help them persevere in populations. Someday this could mean the ability to predict an about-to-be emerging disease before it emerges, as well as strategies for preventing and treating it.

Probiotics: Edible Germs

Probiotics are living microorganisms used as food supplements. The organisms don't necessarily have to colonize the gut, although the aim of consuming probiotics is often to create and maintain healthful gut flora—healthful for the host. These are health benefits beyond the benefits of good nutrition.

Probiotic consumption is most often an attempt to restore the normal microbial gut community after a course of antibiotics has killed off many of them, pathogens and innocents alike. But increasingly probiotics are used to prevent and treat disease.

These health effects are specific to a particular strain of organism and don't necessarily apply to others of the same species.

 Wee Warnings

Probiotics should not be confused with prebiotics. Prebiotics are nonliving food ingredients that selectively stimulate the growth or activity of just one or a few colon bacteria. These bacteria are presumably already there, not introduced. All those prebiotics identified so far are carbohydrates, specifically indigestible polysaccharides often found in vegetables from the onion-garlic family.

The idea is that most gut microorganisms are not just harmless, they are often explicitly helpful. Even though it's not yet clear exactly what they're doing, it's useful to have them down there. The rationale is that they give the normal gut microbes time to build their numbers back up while competing with potential pathogens and keeping them at bay.

Many of us use probiotics unintentionally every time we eat yogurt that contains live cultures of *Lactobacillus*. Yogurt and other fermented foods such as kefir, sauerkraut, and kimchi are probably the chief source of probiotics.

Probiotics are also sold as dietary supplements—meaning that (in contrast to prescription drugs) their health claims have not necessarily been demonstrated. Researchers have even shown that some probiotic products contain no microorganisms at all, despite their claims. A problem for probiotics is that legitimate research on their efficacy is shadowed by the commercialization of probiotic products that may have little scientific support, which makes the claims hard to evaluate.

Nonetheless, research has demonstrated at least some benefits of probiotics. These have been particularly persuasive in those disastrous diarrheal diseases of children. For example:

- Infant formula supplemented with probiotics prevents diarrheal disease in frequently hospitalized children.

- *Lactobacillus rhamnosus* strain GG cures diarrhea in malnourished children, especially those not breast-fed.

- Children with acute gastroenteritis who received *Lactobacillus* probiotics were sick for a shorter time than kids who didn't.

- Probiotics work best in rotavirus-induced diarrhea—a cheering bit of news.

- Probiotics also reduce fecal excretion of rotavirus—meaning, one hopes, a reduction in the transmission of rotavirus diseases.

In addition:

- Live bacteria in yogurt improve the body's handling of lactose (the sugar in milk) in people who are lactose intolerant, so yogurt is a painless way to boost their calcium consumption.

- *Lactobacillus casei GG* in yogurt or probiotic supplements helps mild diarrhea accompanying some cases of *Clostridium difficile* infection.

- *L. rhamnosus GG* was given to pregnant women with a family history of allergic eczema and to their infants for six months, and during two years of follow-up the bacteria-treated babies had much less eczema than untreated ones.

There are many hints that probiotics may have much wider applications too. Scientists are investigating the therapeutic potential of probiotics for treating and preventing infections of wounds, the ears, urinary tract, and skin—and also therapy for cancer, plus reducing cholesterol and preventing tooth decay.

> ### Little Did You Know
>
> Patients who are given broad-spectrum antibiotics already are advised to eat yogurt to reduce the chance of opportunistic colonization by undesirable microbes in the gut as a result of the antibiotic's bactericidal effects on the natural gut flora.

Researchers are hopeful about treating digestive diseases such as irritable bowel syndrome, although to date research results have been equivocal. Both probiotics and prebiotics prevent colon cancer in lab animals, but that effect has not yet been shown in people. Probiotics do, however, reduce fecal activity of enzymes that produce cancer-causing compounds in people.

Livestock producers have already voted in favor. Probiotics are routinely added to animal feeds.

Even if a bug is not a particularly effective probiotic, genetic engineers may make it useful, and we will probably be seeing more of this in the twenty-first century.

For example, researchers have modified a yogurt lactobacillus to release a drug that prevents HIV from infecting cells—essentially a type of passive immunization. As we write, this strategy for delivering anti-HIV drugs is only in the early research stage and has not yet been demonstrated to work in people. But if it pans out, the technique could be an inexpensive and efficient way to dispense medicines and oral vaccines—and it could be used in any facility that can produce yogurt.

CSI: Microbial Forensics

Microbial forensics is a new type of forensic analysis. It investigates crime (including bioterrorism) by studying microbial DNA, and sometimes also proteins or carbs on a microbe's surface. These can serve as a unique "fingerprint" of exactly the kind of microbe used in a crime.

In addition to crime and bioterrorism, microbial forensics can also apply in lawsuits alleging negligence. For example, a patient infected via medical equipment might accuse a hospital of inadequate hygiene, or people felled by food poisoning might sue a restaurant or food packager.

Microbial analysis techniques have been used in the past as a tool of molecular epidemiology, to identify disease organisms—for example in hospital-based outbreaks or food-borne infections. Microbial forensics will be based on similar techniques. The difference is that microbial forensic evidence must persuade not only medical personnel but also judges and juries.

Bolstering the Immune Response

Modifying the host's immune response to respond more effectively, rather than fighting pathogens directly, is advantageous for many reasons when you're trying to fight diseases. For one thing, modulating immune responses is not expected to speed the evolution of resistant microbes, because the modifications don't affect them directly. So it's one solution to the worldwide predicament we're in now, where using antibiotics to cure disease has the dispiriting side effect of creating pathogens that are impervious to them.

In addition, immune system modulators could potentially work against many kinds of pathogens, viruses and fungi, as well as bacteria. They might also be effective as emergency treatment when the exact nature of the pathogen is not yet known—as in the case of a new disease organism or a bioweapon. Moreover, immune system modulators can expand the therapies available to people with weak immune systems.

We are speaking here partly about potential agents for stimulating the immune system. But we also mean agents that damp it down, which might be equally valuable. In many diseases, the injuries wrought by an overenthusiastic immune response are far worse than direct damage from the pathogen. Examples are rheumatic fever following a streptococcal infection and organ damage due to immune system response to hepatitis B virus. Therapies that moderate those destructive overreactions are highly desirable.

The experts agree that no single magic bullet will modify immune system responses against all pathogens in all of those infected. Immune system modulators are more likely to be custom-tailored for specific pathogens, specific groups of patients, or specific points in the progress of a disease—and especially as additions to conventional treatments such as antibiotics and antivirals that go after the pathogen directly.

Little Did You Know

You'll recall we've said the mouth may be the busiest ecosystem in the body and much about it remains to be learned. One recent discovery is mouth bacteria that can get rid of bad breath instead of causing it. They mop up foul-smelling compounds generated by the breakdown of sulfur-containing amino acids in foods. These useful bugs include *Bacillus, Brevibacterium casei, Hyphomicrobium sulfonivorans1, Methylobacterium, Micrococcus luteus,* and *Variovorax paradoxus.* Some of the same odor-eaters work on smelly feet, too.

Passive Immunity

In Chapter 11, we explained passive immunity. Vaccination results in active immunity to a disease by stimulating antibody production. But some viral diseases can also be staved off via passive immunity: direct administration of antibodies made in another host (usually human). Look for more applications of passive immunity in the future.

Passive immunity is a strategy against certain pathogens when no vaccine is available—or when people have just been exposed to a pathogen, or are about to be exposed to one, and their immune systems would not have enough time to make their own antibodies.

The passive immunity approach is not new. People exposed to the bug that causes tetanus have long been given antibodies to the deadly toxin it makes. Immunoglobulins (antibodies, remember?) are also administered to people exposed to hepatitis A, measles, rubella, and rabies, among others. Researchers are working on immunoglobulins against anthrax.

Another approach to preventing anthrax is monoclonal antibodies. Monoclonals are pure antibodies produced from a single cell line and recognizing only a single antigen. Monoclonals are now used against cancer, but there is hope for many other applications, including passive immunity against infectious disease. They might, for example, help deal with antibiotic-resistant pathogens by stimulating cytokines to boost—or reduce—immune response to the diseases they cause.

Researchers are studying passive immunity approaches to potential bioweapons other than anthrax: botulinum toxin, smallpox, tularemia, and plague. Antibodies against Alzheimer's disease are being tested in clinical trials, and a small experimental study that administered immunoglobulin against cytomegalovirus to pregnant women showed that it prevented the disease in most of their babies.

Active Immunity: Vaccines

The active immune system is not being neglected. There are enormous difficulties in developing and producing new vaccines, but there are also new approaches to vaccines.

For example, vaccination can be used as therapy as well as prevention. Rabies vaccine administered soon after infection can keep the disease from developing. Varicella virus vaccine given after chickenpox infection prevents late-life development of herpes zoster (shingles). Look for an expansion of that strategy.

There is also hope for vaccines against chronic infectious diseases such as those caused by protozoa, fungi, and viruses such as HIV. The plan here is to identify antigens that trigger both harmful and helpful immune responses.

Edible Vaccines

The idea behind edible vaccines is to insert human disease protection into food plants. Potatoes, bananas, tomatoes, legumes, rice, and several others are among the possibilities. These genetically modified foods would contain antigens that trigger an immune response but no pathogen genes that could actually cause disease. They could be locally grown and would not need refrigeration, making them less costly to make, transport, and store than conventional vaccines.

In addition, because they would work through the stomach lining, perhaps they would also stimulate the mucosal immune system to produce antibodies, something injected vaccines don't do. That could mean they would work against one of the chief infectious diseases: diarrhea. It might even be possible for a baby to come into the world preprotected if a pregnant woman could consume edible vaccines producing antibodies that she could pass on through the placenta.

However, edible vaccines also face hurdles that don't occur with conventional vaccines. For example, can plants produce antigens that work? If so, can the antigens survive the trip through the stomach and GI tract and still jump-start the immune system? And there might well be political objections too, because the plants that produced vaccines would be genetically engineered.

Virtual Microbial Cells

A virtual microbial cell is an attempt at a computer simulation of a minimal but complete set of genes and metabolic pathways in a single microbial cell. As we write, this hasn't yet been accomplished, but researchers are working on a virtual *Escherichia coli*. *E. coli* was selected because so much (and yet, so little!) is already known about it.

The point is to help researchers build a mock-up of a bacterium or yeast cell that accurately describes all its functions and circuits. They hope that virtual model organisms of this sort will help reveal how the real organisms do things, such as produce energy and evade immune systems. Most of the time they can already predict how fast their *E. coli* will grow, depending on what kind of food it's given.

Many researchers are working on creating a lean and mean *E. coli* that can still function even when equipped with the fewest genes possible. Then they want to model this basic bug on the computer. But when they get to that point, they say, the operations of even the minimal *E. coli* are so complex that no existing computer will be able to simulate them. They must wait a few years for computing power technology to catch up with the complexities of the simplest microbial life.

If these dreams of virtual microbes come true, it should be possible to experiment with simulating different kinds of genetic changes to see how they alter the organism. Virtual microbes may also help with fashioning designer microbes.

Custom-Made Microbes via Synthetic Biology

Synthetic biology is not just a newfangled name for genetic engineering of microbes. It involves not only insertion of new genes, but also attempts to build bacteria from the ground up, bestowing on them abilities and traits never before seen.

Researchers are, for example, designing and building a synthetic version of the *Mycoplasma genitalium* genome. They picked *M. genitalium* because it's the simplest cell known that can replicate on its own. They have divided its tiny chromosome into chunks and then copied each subunit faithfully using chemically synthesized oligonucleotides.

The researchers are devising many versions of each subunit so that when they are combined into chromosomes again, the possible genome arrangements will number in the millions. Each one of these new complete genomes will then be introduced into *M. genitalium* to see whether the bug can function with it—or, intriguingly, if it exhibits any striking new abilities.

A goal of this project is to build and define a minimal cell, one that can survive in the lab with the fewest number of functions. After they have the minimal genome in hand, the researchers hope to build a computer model that can predict cell behavior.

The minimal cell could also be a starting point for adding functions. They envision augmenting the cell with gene pathways that can make organic products for industry, biofuels, or compounds that are hard to make in a pharmaceutical lab.

Other researchers are attempting feats only slightly less ambitious. Some are working on the ribo-regulator, an inserted DNA sequence that integrates into a host microbe's genome, creates RNA that can bind to the ribosome and block (or unblock) production of a particular protein. The goal is to produce a bug that is a factory for making specific proteins. Ribo-regulators are close to commercial availability.

Researchers are dreaming of synthetic microbes that might clean up pollutants, detect chemical weapons, diagnose disease, make hydrogen for fuel out of water and sunlight, and even turn gene therapy into a reliable approach for curing patients. They face immense technical challenges; no one yet knows how far synthetic biology will actually go.

They also worry about its potential for causing damage—either accidental or intentional. What kind of environmental damage could a synthetic bug cause if it escaped the lab? Would it be possible to engineer a pathogen like the Ebola virus to be even more virulent?

Microbiology as a Career

This explosion of knowledge has revealed how much remains to be discovered, and that means plenty of work lies ahead. Some of this work will be in new fields like "green" chemistry, microbial forensics, and biodefense. Some will be in changing ones like bioremediation of pollution and sustainable industrial biotechnology.

Even more will lie in expansion of existing services related to medicine. People who work with pathogens are becoming microbial ecologists, interested not only in the relationship between microbe and host but microbes with other microbes.

People with training in clinical microbiology can work in clinical labs in medical centers or reference labs. But they can also work in public health, marketing, and sales in the pharmaceutical industry and biotechnology. They can teach, and they can do research.

Microbiology training is also good preparation for further education: graduate school, medical school, or veterinary studies. People with a background in this field are also welcome in many allied health professions, such as medical technologists, dental hygienists, medical assistants and physicians' assistants, nursing, and pharmacy.

The Least You Need to Know

- Microbiology is changing its focus from the study of individual microbes to the study of microbial communities and ecosystems.

- Probiotics, edible germs, are living microorganisms used as food supplements, most often to restore the healthful gut ecosystem after a course of antibiotic treatment.

- Microbial forensics, a kind of identification system for microbes, is a new type of forensic analysis used to investigate crime, including bioterrorism, and may also be useful in lawsuits.

- Passive immunity is therapy involving direct administration of antibodies made in another host (usually human). Several passive immunity techniques are already in use, but researchers are working toward others—for example, against bioweapons, Alzheimer's disease, and cytomegalovirus infection of infants.

- Active immunity techniques, such as vaccines, are also the subject of much research, including vaccines used as therapy as well as prevention and vaccines against chronic infectious diseases caused by protozoa, fungi, and viruses.

- Expansion of microbiology suggests that a variety of careers will be possible for people with training in the field.

Glossary of Terms

acquired immune system   another name for the adaptive immune system.

adaptive immune system   the part of the human immune system where the body responds to new pathogens by "learning" their characteristics and developing ways of fighting them off. Adaptive immunity is very specific; it defends against a single kind of invader—sometimes for life. Also called the acquired immune system.

adenine   one of the bases, usually called *A*, that are components of nucleic acids, RNA and DNA. It base-pairs with thymine.

adenosine triphosphate (ATP)   a molecule that stores and transports energy, perhaps the most important product of glycolysis.

aerobic   active only in the presence of oxygen.

aerosol   a mist of tiny particles, no more than a few microns, suspended in air. The particles can be liquid or solid.

aerotolerant anaerobes   microbes that put up with oxygen being present but don't use it.

agar   a polymer made of galactose, used as a solid surface for culturing microbes in a lab. Several different agar mixtures are available commercially.

agbiotech   the biotechnology industries that focus on organisms of agricultural importance.

algae   single-celled eukaryotic microbes that tend to live in water and produce energy via photosynthesis, although there are exceptions. The designation "algae" (or "alga," the singular) is one of convenience, not a reflection of their evolutionary relationships.

amino acid   a building block of proteins.

amoebic dysentery   severe diarrhea, also called amebiasis, an infection caused by a group of protozoans called amoebas, especially *Entamoeba histolytica*.

anaerobic   unable to utilize oxygen.

anion   an atom with a surplus of electrons; it carries a negative electrical charge.

antibiotic   a manufactured drug designed for therapy of particular bacterial infections. Administered to humans or animals, an antibiotic (usually) kills or cripples bacteria only, not other microbes.

antibody   an immune system protein that identifies invading organisms and foreign proteins.

antigen   any substance that can trigger a response from the immune system. As a practical matter, this includes the majority of macromolecules: nearly all proteins and many poly-saccharides as well.

antiseptics   chemicals that interfere with microbial growth and reproduction, crippling or killing them. Antiseptics are used on the skin.

antitoxins   special antibodies that can block the toxins that bacteria produce.

archaea   new group of prokaryotic microbes discovered only in the 1970s.

atoms   the building blocks of molecules.

autoclave   a machine for sterilizing objects using steam under pressure, usually found in microbiology labs.

autoimmune disease   a disease that results from failure of the body's ways of recognizing "self." When one of these mechanisms goes awry, the body interprets its own cells as foreign, and the immune system attacks them. There are many autoimmune diseases, among them multiple sclerosis and type 1 diabetes.

autotroph   an organism that makes its own food by converting inorganic substances—most notably carbon dioxide—into sugar and free oxygen with the help of light energy. The process is called photosynthesis.

bacilli   bacteria that are shaped like rods or sticks. The singular is *bacillus*.

bacteria   a major group of microorganisms lacking an enclosed cell nucleus, and therefore a prokaryote. Bacteria are usually one-celled. Some cause disease in plants, animals, and people. *Bacteria* is plural; the singular form is *bacterium.*

bacterial growth curve   the orderly process of very rapid cell growth in bacteria that microbiologists observe only in their labs.

bactericidal   a substance that kills bacteria. Sometimes called *germicidal.*

bacteriophage   a virus that infects bacteria; also called *phage.*

bacteriostatic   a substance that inhibits bacterial growth and reproduction without necessarily killing the bacteria.

base   a component of the nucleic acids DNA and RNA, made of atoms of carbon, hydrogen, oxygen, and nitrogen in various combinations. The bases are named adenine, thymine (in DNA only), guanine, cytosine, and uracil (in RNA only).

B cells   lymphocytes that remain in the bone marrow until they mature.

Bergey's Manual   the standard reference work on classification of prokaryotes.

binary fission   simple cell division, in which an organism simply splits into two. Fission is a very common nonsexual form of microbial reproduction. Also called mitosis (in eukaryotes), cell division, cell duplication, or replication.

biofilm   a community of organisms attached to a surface. The organisms (mostly microbes) appear to be a layer of slime that forms on rocks, wood, cement, plastic, metal, glass, plant tissue, animal flesh, and nearly any other surface.

biological safety cabinet (BSC)   laboratory equipment specifically designed to contain potentially infectious splashes or aerosols generated in microbiology labs. There are several kinds of BSCs.

biological soil crusts   a desert ecosystem dominated by lichens, cyanobacteria, algae, mosses, and sometimes fungi and bacteria, bound together with soil particles.

biomass   the total weight of a particular amount of organic matter—for example the plant material in a fermenter being converted to ethanol or the microbes in an acre of farmland.

bioreactor   a vessel for carrying out fermentation; a fermenter.

biotechnology   the use of living organisms, usually after genetic engineering of some sort, to make products.

bovine spongiform encephalopathy (BSE)   mad cow disease, a prion disorder.

budding   a form of asexual reproduction, common in some kinds of yeast, in which buds sprout from the yeast cell surface and then break away to form new yeasts.

capsid   a protein matrix enclosing a virus.

cation   an atom lacking electrons; it carries a positive electrical charge.

cell   the basic structural unit of all living things, which does the basic work that keeps an organism going. Both prokaryotes and most eukaryotic microbes consist of just a single cell. Cells absorb food and transform it into energy. They provide structure to the organism and also contain the genetic instruction manual for performing specific tasks.

cell-mediated immunity   the immune system process based mostly on T-cells.

cell wall   a protective semirigid barrier that surrounds a cell outside its membrane. Some microbes, such as some protozoa, do not possess cell walls. While plants, fungi, and most prokaryotes all possess cell walls, the composition and function of their cell walls vary significantly.

chemical bond   a bond that holds atoms together in a molecule; there are several kinds of chemical bonds.

chemoautotrophs   microbes that make their own energy from inorganic (noncarbon) compounds rather than sunlight, a process that assists nitrogen cycling and also reduces pollution.

chlorine   as a gas dissolved in water, the most frequently used method for purifying public water supplies in the United States. It kills algae and bacteria.

chloroplasts   specialized cell structures (organelles) that contain DNA and are also a type of plastids. Chloroplasts contain pigments that capture light for photosynthesis. The best known of these pigments are the green chlorophylls, which give land plants and green algae their color.

cocci   bacteria that are spherical. The singular is coccus.

codon   a group of 3 nucleotides that specifies which one of 20 amino acids should come next in a polypeptide that will form part of a particular protein. More than one codon may refer to the same amino acid.

commensalism   symbiosis between two organisms in which the relationship benefits only one of the symbionts.

complement system   a cascade of dozens of enzymes that augment—complement—the actions of other immune system components. Complement covers microbes with molecules that mark them for swallowing up by phagocytes, it shatters them, it enhances inflammation, and it boosts its own production.

compost   humus, decomposed organic matter that is good for the soil and for plant growth.

conjugation   a form of sexual reproduction in protozoa. Two similar protozoa join together, opening channels between them and exchanging DNA-containing organelles called gametes. Depending on the species, a complicated series of additional cell divisions and exchanges follow, resulting in offspring with a mix of DNA from both original parents. Also the transfer of genetic material in some bacteria by use of a pilus.

covalent bond   a chemical bond in which atoms join together because they are sharing pairs of electrons.

cryptobiotic soil   another term for biological soil crusts.

culture medium   the mixture of essentials that will enable a microbe to survive and increase in an artificial environment such as a lab. Also called a growth medium. Plural: *media*.

cyanobacteria   the organisms that originated photosynthesis. Often wrongly called blue-green algae, they are actually bacteria.

cyst   in microbes, a dormant stage in the life cycle. Cysts are resistant to harsh environments.

cytokine   a protein that is a major communication method between immune system cells.

cytosine   one of the bases, often called C, that are components in the nucleic acids, RNA and DNA. It base-pairs with guanine.

cytoskeleton   a cell structure, similar to scaffolding, which helps cells keep their shape and attaches organelles securely.

defensins   antimicrobial peptides that can kill both gram-positive and gram-negative bacteria, plus other microbes: fungi, protozoa, even viruses.

dehydration   loss of fluids and essential electrolytes.

deoxyribonucleic acid (DNA)   one of the molecules that stores and transfers genetic information.

diatoms   the most common algae, which have shells (exoskeletons) made of silica. Can produce toxic algal blooms.

dinoflagellates   phosphorescent algae that are usually round and protected with stiff body armor made of cellulose and coated with silica. Can produce toxic algal blooms.

disaccharide   a carbohydrate molecule made of two monosaccharides.

disinfectant   a compound that kills microbes on nonliving surfaces (such as a kitchen counter or a surgical instrument) using physical or chemical methods. Disinfectants can sometimes sterilize, but the chief point of disinfection is to reduce the number of microbes.

DNA   see *deoxyribonucleic acid.*

electron   a component of atoms, a particle that carries a negative electrical charge.

elements   substances that cannot be broken down into simpler substances using chemical methods. They include oxygen, nitrogen, hydrogen, and carbon.

emerging disease   according to the National Academy of Sciences, "either a newly recognized, clinically distinct infectious disease, or a known infectious disease whose reported incidence is increasing in a given place or among a specific population."

endocytosis   a form of active transport into eukaryotic cells.

endosymbiosis   a form of symbiosis in which one of the symbiont organisms lives inside the other.

enterotoxin   a protein released by a pathogen in the intestine which causes cramps, nausea, and diarrhea.

enzyme   a protein that helps speed up chemical reactions.

epitopes   unique molecular structures on the antigen surface that are recognized by antibodies. Also, special signature compounds, chemical or protein, attached to proteins used to identify the tagged proteins.

ethylene oxide   a penetrating gas much valued for sterilizing because it can get through packing materials and plastic wraps. It can also sterilize laboratory and hospital equipment that can't be heated. The gas kills microbes and their spores by attacking their proteins.

eubacteria   "true" bacteria, not archaea.

eukaryote   an organism whose genome is organized into a cell nucleus and surrounded by a membrane. Plants and animals are eukaryotes and so are three kinds of microbes: protozoa, fungi, and algae.

extremophile   an organism that can live in an unfriendly environment, very hot or very cold, very acid or very alkaline, etc.

facultative anaerobes   organisms that prefer oxygen, but can live without it.

fermentation   a natural process in which a cell (often a microbial cell) gets the energy it needs by breaking some organic material down into its component parts without using oxygen. Humans can use microbes to ferment specific sugars to create usable end products such as beer and bread. There are several kinds of fermentation.

flagella   tail-like appendages that help microbes move around. Singular: *flagellum*.

flora   the combination of all the microbes that colonize a particular place or host, usually the microbes that are normally there, not sporadic pathogens. When microbiologists say "flora," they very often just mean bacteria.

foodborne disease   illness as a result of consuming food infected with pathogens or their toxins. Not all foodborne disease is food poisoning. Prion diseases acquired via prion-infected meat may not manifest themselves for decades.

food poisoning   gastroenteritis that comes on after eating food containing specific pathogens or the toxins they produce.

foraminifera   also known as forams; protozoans with external skeletons made of calcium carbonate or sand grains glued together.

fungi   a large and diverse group of organisms, many of which (yeasts, molds, mildews) are microbes. Singular: *fungus*.

gamete   a reproductive cell, sperm or egg (ovum), also called a sex cell or germ cell.

gene   a short sequence of genetic material that performs a specific task in a cell. The term usually refers to a structural gene, which issues instructions for making a particular protein or part of a protein.

gene sequencing   discovering the order of bases (nucleotides) in a gene or any other piece of DNA, including the entire genome of an organism.

gene therapy   genetic engineering for treating disease.

genetically modified organism (GM)   a term used to refer to genetically engineered organisms in debates concerning agbiotech and other forms of biotechnology.

genetic engineering   modifying the genes of an organism.

genetic manipulation   modifying the genes of an organism.

genetics the life science that investigates how the genetic material in cells affects what goes on inside them.

genome the full complement of an organism's genetic material, although often the term refers only to nucleic acids in the nucleus.

genomics the study of the full set of an organism's genetic material.

glycolysis the splitting of a glucose molecule to yield energy in the form of ATP. Glycolysis is basic to life on Earth; it is the only metabolic pathway found in every living organism.

gram-negative bacteria bacteria with thin cell walls and a lipopolysaccharide layer outside the walls. They stain with crystal violet but release the stain upon washing with ethanol.

gram-positive bacteria bacteria that have thick cell walls and no membrane outside the walls.

growth medium the mixture of essentials that will enable a microbe to survive and increase in an artificial environment such as a lab. Also called a culture medium. Plural: *media*.

guanine one of the bases, usually called *G*, that are components of the nucleic acids, RNA and DNA. It base-pairs with cytosine.

halophiles salt-loving prokaryotes.

herd immunity resistance of an entire population to disease due to the fact that many of its members are immune (usually because they have been vaccinated against it).

heterotroph an organism that lives on both organic and/or inorganic raw materials collected from the environment. Unlike autotrophs, heterotrophs don't make their own food. Instead, they eat autotrophs—and each other.

human leukocyte antigens (HLA) the human major histocompatibility complex proteins, a component of the immune system.

humoral immunity the immune system process based on antibodies produced by B cells.

humus decomposed organic matter that is good for the soil and for plant growth. Also called *compost*.

hydrogen bond a kind of chemical bond that forms between hydrogen atoms and atoms with a negative electrical charge, often oxygen or nitrogen.

hyphae   fungal structures that typify filamentous fungi, like *Penicillium*. They are string-like in appearance and serve many functions, such as uptake of nutrients, release of enzymes, and reproduction.

icosahedron   a common capsid shape, a polygon with 20 faces.

immune system   the collection of methods an organism uses to distinguish between components that belong to itself and those that enter from outside, called nonself.

immunoglobulin   antibody, abbreviated *Ig*.

inclusion bodies   cell structures used for storage of energy, carbon compounds, and inorganic substances.

inflammation   a distinctive combination of redness, heat, swelling, and pain at the site of an infection. Inflammation helps fight infection but it can also do serious damage to the body.

innate immune system   the immune system that organisms are born with, actually a heterogeneous collection of strategies for battling invading organisms. Innate immunity consists of general methods of coping with a variety of invaders.

interferon   a kind of cytokine. When manufactured in large quantities, used for treating some diseases.

ion   an atom that possesses an unequal number of electrons and protons. Atoms with a surplus of electrons are called anions and carry a negative electrical charge. Atoms lacking electrons are called cations and carry a positive electrical charge.

ionic bond   a kind of chemical bond where atoms are attracted to each other because they possess opposite electrical charges.

junk DNA   the unofficial term for DNA without obvious functions.

leukocyte   white blood cell, a major component of the immune system.

lichen   a symbiotic community of a photosynthetic microbe, usually a green alga or cyanobacterium, with a fungus.

lipid   fat.

lipopolysaccharide   a type of compound, consisting of lipids conjugated with carbohydrates, which decorate the membrane covering the outside of the cell wall in gram-negative bacteria.

lithotrophs   microbes that make their own energy from inorganic (noncarbon) compounds rather than sunlight, a process that assists nitrogen cycling and also reduces pollution. A subtype of heterotrophs.

lymphatic system   an alternative circulatory system, the central hub of the immune system in most animals.

lymphocyte   a white blood cell that is a component of the adaptive immune system. There are two kinds of lymphocytes: T and B cells.

lysis   the breaking open of a cell.

macrophage   a large white blood cell, a phagocyte.

major histocompatibility complex (MHC)   proteins produced by the organism's own genes that help it distinguish self from nonself. Almost all vertebrate cells possess MHC proteins on their surfaces.

membrane   the covering or barrier made up of lipid bilayers that surrounds a cell. It is studded with protein channels and pumps that keep essential molecules inside and govern traffic in and out of the cell.

messenger RNA (mRNA)   the RNA copy of DNA made by RNA polymerase; this is the RNA that gets translated into protein.

methanogens   methane-producing anaerobic archaea.

microaerophiles   microorganisms that grow best in low oxygen concentrations.

microbial oceanography   another term for marine microbiology.

microbiotic soil   another term for biological soil crusts.

μm   a millionth of a meter, usually called a micron and sometimes a micrometer.

mitochondria   an organelle that is the principal source of energy in almost all eukaryotic cells. The singular spelling is *mitochondrion*.

molecule   the fundamental piece of an element, the smallest fragment that still retains the chemical properties of the element.

monoclonal antibodies   pure antibodies produced from a single cell line and recognizing only a single antigen.

monosaccharide   the simplest type of carbohydrate molecule.

MRSA   methicillin-resistant *Staphylococcus aureus*, bacteria that are resistant to beta-lactam antibiotics and are now one of the most serious superbug problems all over the world.

multidrug resistance (MDR)   the ability of some pathogens to defend against several antibiotics at once.

mutation   a change in the DNA sequence.

mutualism   a form of symbiosis in which the relationship between two organisms is beneficial to both partners.

mycelium   a tangle of fungal hyphae.

mycorrhizae   fungi that form a symbiotic association with plant roots and help them absorb minerals and water.

mycoses   diseases caused by fungi.

neutron   a component of atoms, a particle that carries no electrical charge.

neutrophil   an immune system leukocyte that is a type of phagocyte. Also called *polymorphonuclear leukocyte*, *PMN*, and *poly*.

noncoding DNA   DNA that does not code for proteins but whose function is usually unknown.

nosocomial   an infection acquired in a hospital or other healthcare institution, or as a result of medical care. Usually used to mean a hospital-based infection.

nucleic acids   molecules that store and transfer the genetic information that governs the behavior of a cell; DNA and RNA.

nucleoid   cell locale where a prokaryote's DNA resides. Also called nuclear body, nuclear region, and chromatin body.

nucleotide   the basic unit of RNA or DNA. It is made up of three units: a sugar, a base (G, C, A, T, or U), and a phospho group.

oligosaccharide   a carbohydrate made of between three and nine monosaccharides.

oomycete   an organism that resembles a fungus but is not a true fungus. Also called water mold. The water mold *Phytophthora infestans* caused the Irish potato blight.

opportunistic pathogens   microbes that are normally harmless or beneficial, they cause disease only when hosts are weakened by age or other disease and host immune systems are not functioning normally.

organelles   "little organs," specialized cell structures that carry out specific functions. Some organelles originated as symbiotic microbes.

organic molecule   a molecule that contains a carbon atom covalently bonded to one or more hydrogen atoms.

osmosis   the diffusion of liquid from a weak solution to a strong solution across a semipermeable membrane.

osmotic lysis   the bursting of a cell as a result of pressure from osmosis.

parasite   an organism that lives on a host while exploiting and possibly killing it.

parasitism   a form of symbiosis in which one symbiont exploits or injures (and sometimes kills) the other.

pathogen   an organism that causes disease.

peptide   a small protein or piece of a protein.

peptide bond   a kind of chemical bond that holds amino acids together.

peptidoglycan   the complex molecule in a bacterial cell wall that helps make it strong and rigid. Also called *murein* or *PGN*. Many antibiotics keep PGN from being constructed correctly, thus robbing the bug of protection.

pH   a measure of the concentration of hydrogen ions in a fluid, usually water. It is the log_{10} value of the concentration of hydrogen ions in the solution. The measure is expressed on a scale from 0 to 14. A pH of 7 is neutral; anything below 7 is acidic and anything above 7 is alkaline (sometimes called basic).

phage   a virus that infects bacteria; also called *bacteriophage*.

phagocyte   a cell of the immune system that can surround, swallow up, and digest other cells, mostly dead cells and microbes.

phenolics   a broad class of chemical compounds, some of which are disinfectants. They work by denaturing microbial proteins and disrupting their cell membranes.

pheromones   chemical signaling molecules released by an organism to communicate with others, often others of its own species.

photosynthesis   the chemical process inside some living cells that turns the energy from light—generally sunlight—into food, building carbohydrates out of carbon dioxide and water. Most plant cells carry out photosynthesis, and so do cyanobacteria, where the process probably originated.

phycoremediation   the process of using soil algae to remove pollutants like heavy metals and excess nutrients. Also called algal phytoremediation.

phytoplankton   photosynthesizing plankton, mostly cyanobacteria.

pili   hairlike exterior structures that help bacteria stick to surfaces.

plankton   assorted microbes—algae, bacteria, protozoa—free-floating in water, the foundation of the marine food chain. Many of them carry out photosynthesis and are known as phytoplankton.

plasma membrane   the prokaryote cell membrane.

plasmids   tiny free-floating circular pieces of DNA within a bacterium that are separate from its main genome. Plasmids can make copies of themselves and can be transferred to other organisms fairly easily, making them useful for genetic engineering. Genes for resistance to antibiotics are often carried on plasmids, one reason other bacteria can acquire them.

plastids   specialized DNA-containing cell structures (organelles) that perform specific functions, such as chloroplasts for photosynthesis.

polymerase chain reaction (PCR)   a laboratory procedure for making unlimited copies of any piece of DNA less than 15 kilobases long. PCR is probably the single most important methodological development in the life sciences in the past two decades.

polymorphonuclear leukocyte (PMN)   another name for neutrophil, also called *poly*.

polysaccharide   a complex carbohydrate, sometimes consisting of thousands of sugar molecules.

potable   a term for water that is safe to drink.

prion   an oddly folded protein particle capable of causing a group of fatal brain diseases called transmissible spongiform encephalopathies. Prions are counted as microbes, but they are not alive.

Probiotics   living microorganisms used as food supplements.

prokaryote   an organism that does not keep its genetic material in a cell nucleus surrounded by a membrane. Only two kinds of organisms are prokaryotes: bacteria and archaea.

promoter   a stretch of DNA sequence in front of a gene that controls the timing and amount of the gene's expression. In prokaryotes, genes with related functions are sometimes regulated by the same promoter.

protein   an organic molecule made up of units of amino acids. Proteins are essential for the structure, function, and regulation of all cells and tissues in all living things.

protein folding   the process by which proteins form their 3D structure, which governs the way they function.

proteome   the collection of all the proteins in a particular organism. The study of proteomes is called proteomics.

protists   a miscellaneous category of mostly one-celled eukaryotes that are not necessarily closely related; the term is now usually replaced by *protozoa*.

proton   a component of atoms, a particle that carries a positive electrical charge.

protozoa   a simple one-celled eukaryote classified as a microbe. The term is used largely for convenience and does not indicate close evolutionary relationships among members of the group.

quorum sensing   a process by which bacteria communicate with each other by releasing signaling molecules. A single bacterium can perceive the number of other bacteria around it by "measuring" the concentration of signaling molecules. When this concentration reaches a critical mass, the bacteria then can adapt to a change in nutrients, carry out defensive maneuvers, avoid toxins, and (in the case of pathogens) even coordinate their virulence so as to evade the host's immune system.

retrotransposon   a transposable element that requires a reverse transcription step where RNA is used as a template to make DNA.

retrovirus   a family of RNA viruses. Some are pathogens, like HIV, but others are being explored for use in gene therapy.

reverse transcriptase   an enzyme that can copy a piece of RNA into DNA, the reverse of the usual process of using DNA as a template to make RNA.

Rhizobium   a genus of bacteria that live in the roots of plants in the legume family and supply them with nitrogen. Plural: *rhizobia*.

ribonucleic acid (RNA)   one of the molecules that stores and transfers genetic information. It is structurally identical to DNA with two exceptions: (1) the sugar component on RNA contains one more –OH group and (2) instead of the thymine base, RNA is made with uracil base.

ribosomal RNA (rRNA)   the RNA that helps build ribosomes for making a protein. The sequence of rRNA is often used to determine evolutionary distance between organisms.

ribosome   a complex cell structure for producing proteins. Ribosomes are made of proteins plus a particular kind of RNA called ribosomal RNA (rRNA). See also *translation*.

RNA   see *ribonucleic acid*.

RNA polymerase (RNAP)   the enzyme that makes RNA molecules using DNA sequences as templates. See also *transcription*.

rusts   a group of plant diseases caused by fungi.

saprophyte   an organism that eats only dead material.

spirilla   bacteria that are squiggly but rigid like a spiral or helix.

spirochete   a bacterium that is spiral-shaped but bendable.

spore   a reproductive structure that is similar to a seed except that it doesn't contain stored food like a seed. Many microbes form spores. Spores tend to be durable and can survive for a long time in the environment before developing into microbes.

sporozoans   a group of protozoans, sometimes called *Apicomplexa*, which are nearly all parasites and are mobile although they lack flagella, pseudopods, or cilia.

sterilization   a process that exterminates all forms of life completely, no exceptions. A sterile surface by definition harbors no microbes of any kind, nor spores that could germinate into microbes.

superantigen   a toxin produced by bacteria that latches on to antigen receptors on T cells, triggering production of cytokines that shut down immune responses.

superbug   a bacterium that is resistant to several antibiotics.

symbiont   an organism that lives in or on another organism but does it no harm and sometimes even benefits the host.

symbiosis   an intimate association between two (or sometimes more) different kinds of organisms. Often a community of microbes, or a community of microbes and other life forms.

T cells   lymphocytes that mature in the thymus.

thermoacidophiles   aerobic archaea that are acid- and heat-loving.

thymine   one of the bases, usually called *T*, that are components of the nucleic acid DNA. It base-pairs with adenine.

transcription   the process in which the information coded in DNA is copied (transcribed) into a molecule of RNA. The ladder of DNA splits in two, and a half-ladder of DNA serves as a kind of mold or pattern for making an RNA copy.

transduction   a method of DNA acquisition in bacteria in which foreign genes are carried into a bacterium by a virus called a phage. This can occur either in nature or in a lab, where a scientist employs transduction to genetically engineer a bacterium.

transfer RNA (tRNA)   the RNA that carries amino acids to the ribosome where a protein is under construction.

transformation   a method for DNA acquisition in which a bacterium takes up pieces of DNA from outside sources and incorporates it into its own DNA. These sources can include related species, plasmids, viral genomes, or even genomic bits that dead bacteria have left lying around.

translation the process in which proteins are made from the deciphering of information encoded in mRNA.

transposable element transposon.

transposon a transposable element, a chunk of DNA that can move around in a cell, often simply jumping to another part of the same chromosome or to a plasmid.

uracil one of the bases, usually called *U*, that are components in the nucleic acid RNA. It base-pairs with adenine.

vaccination administration of a harmless or crippled version of a pathogen, or a piece of a pathogen, or a toxin the pathogen produces, to stimulate a long-term immune response without making the host actively ill.

vaccine a preparation of particular antigens that activates the immune system without producing disease. The point is to persuade the immune system to mount a defense—to make antibodies—against specific pathogens, a defense that will be ready if infection does occur. Vaccines are often administered by injection, but some are given by mouth.

vacuole a cell structure or organelle that digests nutrients and sometimes also has other functions.

vector the word *vector* is used in two ways in this book. In connection with disease, a vector is an organism, often an insect or rodent, which carries a pathogen and can transmit it to new hosts. In genetic engineering, a vector is a plasmid, virus, or bacterium that is used to transport a new gene or genes into a host cell.

vibrio bacteria that are shaped like commas.

virion a complete virus particle before it enters a cell.

virology the study of viruses.

virus an infectious particle that can grow and reproduce only inside a host cell; unlike other microbes, it is not alive.

water mold see *oomycete*.

yeast a kind of fungus that (unlike most fungi, which are aerobes) gets its energy by converting sugar into carbon dioxide and ethanol (alcohol) in an anaerobic process called fermentation.

zoonosis an animal infection that can be transmitted to humans. Plural: *zoonoses*. The majority of human infectious diseases are zoonoses.

How to Handle Microbes

We don't mean to be alarmist here, far from it. The vast majority of microbes are our friends—or at least not our enemies. A good thing, since we spend our entire lives swimming in a sea of them. Microbes are unavoidable and in most cases don't even need to be avoided.

But much of this book has been about microbes that do not fall into the friendly category. Some are dangerous or even lethal and to be shunned if you can. And, as we have pointed out more than once, many microbes that are normally harmless can turn vicious if conditions are right.

Safety in the Microbiology Lab

Over many decades, thousands of lab workers have gotten infections on the job. There have even been deaths. Marburg fever, you'll recall, was first identified after it felled European lab employees who had been handling blood and tissue from African monkeys.

Marburg is a special case, since it involved a previously unknown disease. Most lab-related illnesses have come from contact with more everyday infectious organisms. Many of these infections could have been avoided with simple precautions that are standard in microbiology labs—or even just plain hand washing.

Detailed manuals on lab safety are widely available; you'll find some URLs in the list of resources in the back of this book. But a good many lab safety measures come under the heading of just plain common sense. For example:

◆ Wash your hands thoroughly before and after your session.

◆ Do not eat or drink (or smoke, of course) in the lab.

◆ Use protection—safety glasses, gloves, lab coats (especially if you're wearing shorts, tank tops, or other skin-exposing clothing).

◆ Stow your backpack and other gear well away from the work area.

◆ Disinfect bench tops before and after your session.

◆ No pipetting by mouth, ever!

◆ Don't take equipment (or cultures!) out of the lab.

◆ Notify the supervisor immediately about spills, cuts, and other accidents.

◆ Dispose of or sterilize used equipment as instructed.

◆ In medical labs where potentially infectious blood and body fluids are handled, special precautions may apply, and generally will be posted prominently.

Tiny Tips

We've said it before, and we'll say it again: wash your hands! Here's how: wet your hands with clean warm water. Use soap and scrub all your hand surfaces for at least 20 seconds. Don't guesstimate, count to 20 slowly. Get under your fingernails too. Rinse really well. Dry with paper towel or a dryer. No soap and no water? An alcohol-based gel from the drug store is an acceptable substitute in most situations.

Aerosols and Biological Safety Cabinets

In particular, avoid breathing in aerosols. An aerosol is a mist of tiny particles, no more than a few microns, suspended in air. The particles can be liquid or solid. In a lab, aerosols are sometimes caused by accidents such as spills. But they also can be due to, say, removing the cover from a culture-containing glassware that has just been shaken.

Inhaling an infectious aerosol is one of the most frequent causes of lab-acquired disease. Tasks that might generate an aerosol are best done in a biological safety cabinet. The biological safety cabinet (BSC) is specifically designed to contain potentially infectious splashes or aerosols that are inevitably generated in microbiology labs.

There are three main kinds. Class I and Class II BSCs are negative-pressure ventilated open-front cabinets but still provide lab workers—and their surroundings—with high-quality protection. Both Class I and Class II BSCs are regarded as good containment systems even for some high-risk microbes if they are used properly.

The highest containment level for working with the most dangerous microbes is provided by Class III BSCs, which are totally enclosed ventilated cabinets that are gas-tight. They have portholes with attached rubber gloves, and all procedures are performed through the portholes with those gloves. Associated equipment like incubators and refrigerators must be interconnected, and a Class III BSC also must be connected to an autoclave for sterilizing.

Biosafety Levels in Laboratories

There are four laboratory biosafety levels, each of which combines lab practices, methods, equipment, and facilities that are appropriate for the kinds of tasks to be performed in that lab. They take into account factors such as how an infectious microbe is transmitted.

Biosafety level 1 (BSL-1) is designed for high school and college labs and also for professional labs working with well-characterized microbes that do not normally cause disease in healthy adult people. Some examples are *Bacillus subtilis, Naegleria gruberi,* and canine hepatitis virus. These labs are assumed to need no special precautions other than the commonsensical ones we outlined above.

Biosafety level 2 (BSL-2) is designed for work with microbes that pose "moderate" risks. These are defined as agents that are present in the surrounding community and also can cause human disease. Some examples: hepatitis B virus, HIV, *Salmonella, Toxoplasma.* The organisms will usually not be transmitted via aerosol.

BSL-2 work includes investigations of human blood, body fluids, tissues—or cell lines that might or might not contain an infectious agent. It goes on in clinical, diagnostic, and some teaching labs, but the microbes can be contained with good standard microbiological techniques. In these labs, the likeliest hazards are accidents that involve the skin or mucous membranes—or inadvertent consumption of a microbe. Another possibility is a needle stick or wound with another sharp instrument. Workers in these labs should use BSCs and protective gear where appropriate.

Biosafety level 3 (BSL-3) labs are those where the pathogens are dangerous, possibly lethal, and where they might be transmitted by breathing them in. Examples include *Mycobacterium tuberculosis, Coxiella burnetii,* and St. Louis encephalitis virus. The chief risks in a BSL-3 lab could be inadvertent consumption of a microbe or breathing in infectious aerosols.

In BSL-3 labs there is a lot of emphasis on protecting lab workers—and people outside the lab—from aerosols. It's routine, for example, that all microbe handling is done in a BSC. Access to the lab may be controlled, and it may have special ventilation systems to keep bugs from escaping to the outside world.

Biosafety level 4 (BSL-4) labs provide the highest containment. BSL-4 labs are where research on the most dangerous microbes happens—dangerous not only because they lead to fatal diseases but also because there is no vaccine or treatment for them. These microbes pose a high risk for lab workers. One example is the Marburg fever virus we cited earlier.

BSL-4 labs are where the workers labor exclusively in Class III BSCs or wear Moon Suits (also known as full-body air-supplied positive-pressure personnel suits). The BSL-4 lab is usually a separate building or otherwise completely isolated and has completely separate ventilation and waste-disposal systems. There aren't very many BSL-4 labs in the world. As we write there are only two in the United States, although more have been proposed in light of worries about bioterrorism.

How to Grow Microbes

Provide the Basics

The field of microbiology depends on being able to grow and study microbes in the lab. How is that done?

Microbes consist mostly of only a few chemical elements: carbon, oxygen, hydrogen, and nitrogen, plus potassium, sulfur, iron, calcium, magnesium, and phosphorus—and some trace elements, vitamins, and growth factors that vary depending on the microbe. In order to grow a microbe in a lab (that is, increase its numbers), you must supply it with a carefully balanced mixture of these basics, and occasionally other things as well. An inadequate supply of any one of the essentials will limit your microbe's growth.

The mixture of essentials that will enable the microbe to survive and increase in an artificial environment is called a growth medium or culture medium (plural: *media*). The conditions differ most dramatically between those needed to grow microbes, mostly single cells, and those needed for culturing cells from more complex organisms.

Types of Growth Media

We have noted several times that many microbes are difficult to grow in the lab, although new methods for culturing particularly fussy microbes are developed all the time. Still, researchers have come up with several standard mixtures for growing, transporting, and storing many different kinds of organisms and cells. These growth media are available commercially, although anyone who wants to take the trouble can make most of them from scratch using standard formulae.

There are several different types of culture media:

♦ Defined media are those in which all the constituents are known and there are often only a few of them

♦ Complex media contain some components that are unknown

♦ Selective media encourage growth of a particular kind of microbe, for example gram-positive or gram-negative

♦ Differential media permit growth of more than one kind of microbe but contain components that make it possible to distinguish one from the other

Defined media are important research tools because investigators often need to know exactly what elements the microbe has available for use.

Complex media are important because they are versatile and can serve for more than one kind of microbe—including microbes whose precise requirements are not yet known.

Complex media often include sugars, plus peptones, which are bits of partially digested proteins from meat, gelatin, or other proteins and yield carbon and nitrogen. They may contain extracts of lean beef or brewer's yeast, which also are sources of carbon and nitrogen—plus amino acids, vitamins, nucleotides, and minerals.

Selective media contain substances (often dyes) that inhibit one kind of microorganism and so encourage the growth of others. The dyes crystal violet and fuchsine suppress gram-positive bugs but not gram-negative ones, permitting the latter to grow freely. Media used to detect contamination by *Escherichia coli*, a gram-negative bug, contain dyes that inhibit gram-positives. Other selective media contain nutrients that only one kind of microbe can use, which permits researchers to detect it in a sample.

Differential media help researchers tell bugs apart with the help of components like sugars, dyes, or blood. A blood-containing medium, for example, can distinguish between hemolytic bugs (bacteria like strep and staph that destroy red blood cells) and nonhemolytic ones. Colonies of hemolytic bacteria are surrounded by clear areas because the red cells have been wiped out.

Solid Culture and Agar

Microbes can be grown in a nutrient broth and often are. But in broth it's difficult to isolate specific microbes so that they can be grown into pure cultures from a single cell. Sometimes it's best to culture microbes on a sterile solid surface.

Here's the way it's done: the surface is inoculated with microbes, which are then "streaked" over the surface in order to separate out a few individual cells. Each colony that develops from one of these cells will be a pure culture. Microbiologists can often identify a microbe by the characteristic appearance of the colony.

The main tool for providing this solid surface is agar. Agar, a polymer made of galactose, is a component of the cell walls of red algae or seaweed. Agar is a bit like gelatin: it dissolves in hot water and then gels as it cools to form a solid surface good for culturing microbes. Most microbes can't decompose agar.

Like other growth media, powdered agar is available from commercial sources, often in combination with other useful components. There are many different agar mixtures, depending on what information is being sought. Following are three examples:

◆ Blood agar contains animal blood cells and can be used, as described above, for identifying hemolytic organisms.

◆ The popular MacConkey's agar favors the growth of gram-negative bugs. It contains crystal-violet dye (which inhibits gram-positives) and also bile salts, lactose, a red dye that stains microbes that ferment lactose, and peptone.

◆ Neomycin agar contains the antibiotic neomycin, so the culture will grow only microbes that are resistant to it.

Counting Microbes

It's often desirable to know just how many microbes there are in a particular sample—for example, when testing swimming water for the coliforms that may mean it's not clean enough. There are a number of ways to count organisms. We describe four: plate count, microscope count, membrane filtration, and estimation of most probable number.

The plate count is just what it sounds like: counting the number of bacterial colonies that grow up from a diluted sample growing on a Petri dish or plate. Multiplying the number of colonies by the amount the sample was diluted will give a reasonable estimate of the number of microbes in the original sample.

Microbes can also be counted with a microscope and counting chamber, which is a special kind of microscope slide etched to a specific depth to hold a particular volume of organisms. There are different kinds of counting chambers for prokaryotes and eukaryotes.

The membrane filtration method is a way of determining which microbes are present in water. Microbes are trapped by passing the water sample through a filter. The microbe-laden filter is then placed on differential growth media, and the resulting growth of colonies reveals which microbes are in the water.

The "most probable number" (MPN) is a statistical estimate (rather than a count) of the number of microbes in a sample. It is based on the number of positive results from a series of liquid culture tubes. MPN is generally expressed as the number of microbes per 100 ml of water.

Resources

You can find tons of information on microbes on the Internet. We've included a few sites sponsored by the government and professional scientific organizations. We have not included academic sites (*.edu) or commercial sites (*.com) because there are too many. But there are plenty of excellent ones that you can find with a brief search. The sites listed in this appendix also contain links to additional information on microbial topics.

General Sources on Microbes

American Society for Microbiology
www.asm.org/

Microbe World
www.microbeworld.org/

USDA Microbiology Laboratory Guidebook
www.fsis.usda.gov/Science/
Microbiological_Lab_Guidebook/
index.asp

Wikipedia is a free encyclopedia that is an excellent source on several microbe-related topics. The scientific information is generally sound. But some entries are more reliable than others, so Wikipedia material should be read in conjunction with other sources.
www.wikipedia.org

Bacteria

National Institute of Allergy and Infectious Diseases
www3.niaid.nih.gov/

Virtual Museum of Bacteria
www.bacteriamuseum.org/

Bioweapons

The National Institute for Allergy and Infectious Diseases at the U.S. National Institutes of Health.
www3.niaid.nih.gov/biodefense/

The U.S. Centers for Disease Control and Prevention.
www.bt.cdc.gov/index.asp

Disease

Centers for Disease Control and Prevention
www.cdc.gov/

Food and Drug Administration
www.fda.gov/

Food

The FDA's wonderful Bad Bug Book
vm.cfsan.fda.gov/~mow/intro.html

USDA Food Safety and Inspection
www.fsis.usda.gov/

Fungi

Ohio State University introduction to fungi.
www.hcs.ohio-state.edu/hcs300/fungi.htm

Mycological Society of America
www.msafungi.org/

Genetics and Genome Projects

The U.S. Department of Energy is the government's sponsor of microbial genomics.
doegenomes.org/

Primer on microbial genomics
microbialgenome.org/primer/primer.shtml

Safety

CDC biosafety info
www.cdc.gov/od/ohs/biosfty/biosfty.htm

More safety info at
www.cdc.gov/od/ohs/hslinks.htm

American Biological Safety Association
www.absa.org/index.shtml

Soil

USDA's Soil Primer
soils.usda.gov/sqi/concepts/soil_biology/index.html

Viruses

All the Virology on the WWW
www.virology.net/

Index

Numbers

5-fluorocytosine, 150

A

AAV (adeno-associated virus), 78
abnormal proteins, 66
acetic acid, 267
Acetobacter, 22
acne vulgaris, 164
acquired immune systems, 120, 124
  lymphocytes, 125-126
    cell-mediated immunity, 126
    humoral immunity, 126-128
  memory, 125, 128
acquired resistance, 151
actinomycetes, 16
active immunity, 285-286
active transport, 94-95
acyclovir, 149
adaptation, microbes, 7
adaptive immune systems, 120, 124
  lymphocytes, 125-126
    cell-mediated immunity, 126
    humoral immunity, 126-128
  memory, 125, 128
adenine, 59
adeno-associated virus (AAV), 78

adenovirus, 187
adenoviruses, 41, 78
aerobes, 10
aerobic bacteria, 15
aerotolerant anaerobes, 15
African sleeping sickness, 29, 74, 223-224
AIDS, 42, 141, 158, 175, 190-193
  avoiding, 132
alcohol fermentation, 267
alcohols, as disinfectants, 138
algae, 4, 8, 30, 231
  algal blooms, 33
  blue-green algae, 31
  brown algae, 31
  detriments of, 33-34
  fragmentation, 32
  green algae, 31
  marine food chain, role in, 32-33
  photosynthesis, 30-31
  red algae, 31
  Red tides, 33
  spores, 32
  types, 31-32
algal blooms, 33
allergies, 129
  fungi, 37, 229
alpha amino acid, 57
Alzheimer's disease, 47, 247
amantadine, 149
amebiasis, 132
amino acids, 56-57, 66
amnesic shellfish poisoning, 34
amoebas, 29
amoebic dysentery, 28

amphotericin, 150
amphotericin B, 150
ampicillin, 146
amyotrophic lateral sclerosis, 247
anaerobes, 10, 22
anaerobic bacteria, 15
anaphylactic shock, 129
animals
  diseases, 242-243
  viral infections, 45-46
anions, 53
Anopheles gambiae, 217
anthrax, 204, 244, 276
  bioterrorism, 274-275
antibiotics, 11, 22, 42, 132-133, 145-148
  beta-lactam antibiotics, 146
  combinations, 148
  fungi, 36
  molds, 146-147
  mycins, 147
  myxobacteria, 19
  new discoveries, 155-156
  resistance, 150-155
  side effects, 148
  sulfa drugs, 147
  tetracyclines, 147
antibodies, 65, 125-127
  humoral immunity, 126-128
  Y-shaped antibodies, 126-127
antifungal drugs, 150
antifungals, 132
antigens, immune systems, 120

antimicrobials, 132
antiparasitic drugs, 132
antiseptics, 132, 139-140
antiviral drugs, 132, 149-150
Apicomplexa, 29
Apollo 12 mission, 18
arboviruses, 180
   bunyaviruses, 183
   dengue fever, 181-182
   rubella, 183
   West Nile virus, 182-183
   yellow fever, 182
archaea, 9, 13-14
   extremophiles, 23
   halophiles, 23
   methanogens, 23
   scientific research of, 23
   shapes, 23
   thermoacidophiles, 23
archaeal cells, 89-90
archaebacteria, 22
arenaviruses, 183
Armillaria ostoyae, 35
aspergillosis, 230
Aspergillus flavus, 242
asthma, 124, 229
atherosclerosis, 124
athlete's foot, 37
atoms, 52
   chemical bonds, 52-53
autoclaves, 135
autoclaving, 134
autoimmune diseases, 129,
   174, 177
autotrophs, 10, 15-17, 30
avian influenza, 244
avian viruses, 140
avoparcin, 152
Azotobacter, 109
AZT, 149

B

B cells, 125-126
   humoral immunity,
   126-128
B virus, 194
B19 parvovirus, 188
bacilli, 16
Bacillus, 204-205
Bacillus anthracis, 276
Bacillus megaterium, 16
Bacillus subtilis, 268
Bacillus thuringiensis, 22
bacitracin, 147
bacteria, 4, 9, 13, 22
   aerobic bacteria, 15
   anaerobic bacteria, 15
   antibiotics, 145
      molds, 146-147
      mycins, 147-148
      new discoveries,
      155-156
      resistance to, 150-155
      side effects, 148
      sulfa drugs, 147
      tetracyclines, 147
   autotrophs, 15-17
   bacterial growth curve,
   97-99
   contents, 17-19
   diseases, 21, 197-199
      Bacillus, 204-205
      Bordetella, 210
      Brucella, 208
      Chlamydia trachomatis,
      211
      Clostridium, 206
      *Corynebacterium dip-
      theriae*, 210
      Haemophilus, 209
      Legionella, 213

Mycobacterium, 210-211
Mycoplasma, 207
Neisseria gonorrhoeae,
   203
normal anaerobic flora,
   205-206
Pseudomonas, 207-208
Rickettsiae, 213
Staphylococcus, 199-200
Streptococcus, 201-202
Treponema pallidum,
   212-213
Yersinia pestis, 208-209
energy, derivations, 15
eubacteria, 14
extremophiles, 18
free-living bacteria, 18
gram-negative bacteria,
   15-16
gram-positive bacteria,
   15-16
green sulfur bacteria, 20
groupings, 19-20
heterotrophs, 15-17
identifying, 17
MDR bacteria, 151
mutualism, animals,
   110-111
myxobacteria, 19
nanobacteria, 17
normal human flora
   gut microbes, 158-163
   mouth microbes,
   166-169
   respiratory tract
   microbes, 169
   skin microbes, 163-165
   urogenital tract
   microbes, 170
oxygen, relationship with,
   15
photosynthesis, 15

plasmids, 18
prevalence of, 14
quorum sensing, 19-20
reproduction, 18
resiliency of, 18
shapes, 16
typical structure, 14
viral infections, 43-44
bacteria, importance of, 15
bacterial cells, 85-89
bacterial growth curve, 97-99
bactericidal, 133
bacteriophages, 44
bacteriostatic, 133
Bacteroides thetaiotaomicron, 159
barrier methods, innate
immune systems, 121
bases
nucleotides, 59
RNA, 67
beer, production of, 21
Bergey's Manual, 17
beta-lactam antibiotics, 146
beta-lactamases, 151
bifidobacterium, 163
bile duct obstruction, 177
binary fission, 18
biodegradable plastics, 11
biofilms, 106-108, 133
biological soil crusts, 262
biological warfare, 273-277
bioluminescence, 21
biotechnology, 265-266
controversies, 272-273
environmental biotechnol-
ogy, 268-270
fermentation, 266-268
industrial biotechnology,
270-272
medicinal biotechnology,
270
bioterrorism, 12, 273-277

bird flu, 244
bird viruses, 140
Black Death, 134, 208
Blastomyces dermatitidis, 230
blue-green algae, 31
bodily fluids, enzymes, 121
boiling as sterilization, 134
Bordetella, 210
Borellia burgdorferi, 213
Botrytis cinerea, 242
botulism, 238, 276
bovine spongiform
encephalopathy (BSE), 47
brown algae, 31
Brucella, 208, 276
brucellosis, 276
bubonic plague, 11, 134,
208-209
Buckminster Fuller geodesic
domes, 41
bunyaviruses, 183
Burkholderia, 276
Burkitt's lymphoma, 247

C

Campylobacter jejuni, 235-236
cancer, 42
Candida, 141, 150, 227-229
Candida albicans, 37, 102
candidiasis, 147, 227-229
capsids, 40
carbohydrates, 55
carbon, 52
carbon dioxide, 54-55
carboxylic acid, 57
carcinogenic viruses
herpesviruses, 193-195
papillomaviruses, 193
cardiovascular diseases,
microbes, 124, 248-249
careers, microbiology,
287-288

caries, 167
cat-scratch fever, 20
catalysis, enzymes, 65
cations, 53
cavities, 167
cell-mediated immunity, 126
cells, 81
archaeal cells, 89-90
B cells, 125
humoral immunity,
126-128
bacterial cells, 85-89
microbial cells
commonalities, 82-84
eukaryote cells, 90-91
membranes, 84
structure, 82-85
walls, 84
mitochondria, 115-117
proteins, 65
T cells, 125
cell-mediated immunity,
126
virtual microbial cells, 286
white blood cells, 122-123
cephalosporins, 146
Chagas's disease, 74, 223
cheese, production of, 21
chemical bonds
atoms, 52-53
peptide bond, 57
chemical sterilization, 136
chestnut blight, 242
Chlamydia, 18, 141, 248
Chlamydia pneumoniae, 247
Chlamydia psittaci, 276
Chlamydia trachomatis, 211
Chlorobium, 20
chloroplasts, 31, 115
cholera, 44, 132, 244
chronic emerging diseases,
247-249

ciliates, 29
Cipro, 146
cirrhosis, 178
classifications
   protozoa, 29
   viruses, 40
clean water, achieving,
   257-259
Clostridium, 206
Clostridium botulinum, 238, 276
Clostridium difficile, 153-154,
   282
Clostridium perfringens, 276
cocci, 16
Coccidioidomycosis, 231
cofactors, 47
Colletotrichum magna, 112
commensalism, 111-112
commensals, normal human
   gut microbes, 158-163
   mouth microbes, 166-169
   respiratory tract microbes,
      169
   skin microbes, 163-165
   urogenital tract microbes,
      170
common colds, 42, 175
complement, 124
compost, microbes, 263-264
conformations, 58
congenital mental retardation,
   247
conjugation, microbes, 99
constant region (Y-shaped
   antibodies), 127
consumption, 210
controversies, biotechnology,
   272-273
coronaviruses, 186-187
Corynebacterium diptheriae,
   210
covalent bonds, atoms, 53
cowpox virus, 40

Coxiella burnetii, 276
Coxsackie viruses, 176
Creutzfeldt-Jakob disease
   (CJD), 48
Crohn's disease, 129
Cryphonectria parasitica, 242
cryptococcal meningitis, 150
Cryptosporidium, 277
custom-made microbes,
   286-287
cyanobacteria, 9, 15, 31-33
cyanobacterium, 31
cysts, protozoa, 28
cytokines, 123, 144
cytomegalovirus, 141, 194,
   247
cytomegalovirus infection,
   195
cytoplasm, 82
cytosine, 59

D

Darwin, Charles, 223
Death phase (bacterial growth
   curve), 98
defensins, 121
Deinococcus radiodurans, 21
dementia, 191
dengue, 141, 181-182, 244
dental plaque, 166
dermafflictions, fungi,
   229-230
dermatitidis, 230
designer genes, 78-79
diabetes, 247
diarrhea, 11, 20, 219-221
diffusion, 94-95
dinoflagellate toxins, 33-34
diphtheria, 210, 244, 274
disaccharides, 55
diseases, 10

African sleeping sickness,
   74
Alzheimer's disease, 47
amnesic shellfish poison-
   ing, 34
amoebic dysentery, 28
animals, 242-243
athlete's foot, 37
autoimmune diseases, 174
bacterial diseases, 21,
   197-199
   Bacillus, 204-205
   Bordetella, 210
   Brucella, 208
   Chlamydia trachomatis,
      211
   Clostridium, 206
   *Corynebacterium dip-
      theriae*, 210
   Haemophilus, 209
   Legionella, 213
   Mycobacterium, 210-211
   Mycoplasma, 207
   Neisseria gonorrhoeae,
      203
   normal anaerobic flora,
      205-206
   Pseudomonas, 207-208
   Rickettsiae, 213
   Staphylococcus, 199-200
   Streptococcus, 201-202
   Treponema pallidum,
      212-213
   Yersinia pestis, 208-209
Chagas's disease, 74
Creutzfeldt-Jakob disease
   (CJD), 48
curing
   antibiotics, 145-148,
      155-156
   antifungal drugs, 150
   antiviral drugs, 149-150
   resistance, 150-155

emerging diseases, 243-245
  chronic diseases,
    247-249
  Ebola, 245-246
  Marburg fever, 245-246
foodborne diseases,
    234-235
  Campylobacter jejuni,
    235-236
  Clostridium botulinum,
    238
  Escherichia coli, 236-237
  food preservation,
    240-241
  Salmonella typhi, 236
  Shigella, 237-238
  spoilage, 238-240
fungal diseases, 226-227
  allergies, 229
  aspergillosis, 230
  asthma, 229
  Blastomyces dermatitidis,
    230
  Candida, 227-229
  candidiasis, 227-229
  Coccidioidomycosis, 231
  dermafflictions, 229-230
  Histoplasma capsulatum,
    231
  mycotoxins, 229
  Pneumocystis jiroveci, 227
fungi, 35, 37
immune systems, 120-121,
    129-130
jock itch, 37
kuru, 46
Legionnaires' disease, 30
leishmaniasis, 74
mad cow disease, 47
malaria, 29
microbes, 7
parasitic diseases, 112
plants, 241-242

pneumocystis pneumonia,
    35
prevention, 131-133,
    143-144
  antiseptics, 139-140
  disinfectants, 137-139
  sterilization, 133-137
  vaccines, 140-143
prions, 46-47
protozoa, 28-30
protozoan diseases, 215,
    225-226
  gastroenteritis, 219-221
  Leishmania, 224
  malaria, 216-219
  Toxoplasma gondii, 225
  Trichomonas vaginalis,
    221-222
  Trypanosoma brucei,
    223-224
  Trypanosoma cruzi,
    222-223
SCID (severe immune
  deficiency), 78
smallpox, 40
  eradication of, 43
tobacco mosaic virus
  (TMV), 40
toenail fungus, 35
viruses, 40-46, 174-175
  adenovirus, 187
  arenaviruses, 183
  coronaviruses, 186-187
  hepatovirus, 177-179
  herpesviruses, 193-195
  influenza viruses,
    184-185
  insect-borne viruses,
    180-184
  norovirus, 190
  papillomaviruses, 193
  paramyxoviruses, 186
  parvovirus, 187-188

picornaviruses, 175-176
  poxviruses, 195
  rabies, 184
  retroviruses, 190-193
  rotavirus, 188-189
yeast infections, 37
disinfectants, 132, 137
  alcohols, 138
  heavy metals, 138
  hydrogen peroxide, 138
  phenol, 139
  phenolics, 139
  surfactants, 138
DNA (deoxyribonucleic acid),
    59-61, 64
  genes, 66
  genetic sequencing, 73-75
  junk DNA, 71-72
  mutations, 66
  proteins, 67-73
  RNA, compared, 67-73
  transcription, 68-69
  transposons, 72-73
Domagk, Gerhard, 147
double-stranded DNA, 59
drugs
  antibiotics, 145
    combinations, 148
    molds, 146-147
    mycins, 147
    new discoveries,
      155-156
    resistance, 150-155
    side effects, 148
    sulfa drugs, 147
    tetracyclines, 147
  antifungal drugs, 150
  antiviral drugs, 149-150
  medicinal biotechnology,
    270
Duchesne, Ernest, 146
dysentery, contracting, 28

E

E. coli. See Escherichia coli 0157:H7
Ebola virus, 67, 244-246
edible vaccinations, 285-286
electrons, 52
elements, microbes, 51-54
elements, molecules, 52
elephantiasis, 114
emerging diseases, 243-245
  chronic diseases, 247-249
  Ebola, 245-246
  Marburg fever, 245-246
encephalitis, 182, 277
encephalopathy, 47
endocytosis, 122
endogenous retroviruses, 190-191
endosymbiosis, 106
  organelles, 114-118
  secondary endosymbiosis, 117
energy, bacteria, derivations, 15
Entamoeba histolytica, 28, 220-221
enterococci, 152
Enterococcus, 141, 146
Enterohemorrhagic E. coli (EHEC), 237
Enteroinvasive E. coli (EIEC), 237
Enteropathogenic E. coli (EPEC), 237
Enterotoxigenic E. coli (ETEC), 237
enteroviruses, 176, 244, 247
envelopes, viruses, 40
environmental biotechnology, 268-270

enzymes, 11, 56, 65-67
  bodily fluids, 121
  catalysis, 65
  complement, 124
Epsilon, 276
Epstein-Barr virus, 194-195, 247
ergosterol, 150
erythromycin, 147
Escherichia coli 0157:H7, 16, 41, 141, 147, 234-237, 244, 276
  quorum sensing, 20
ethanol, 11
eubacteria, 14
Euglena, 29
eukaryotes, 8, 25
  algae, 30-34
    detriments of, 33-34
    marine food chain
  cells, 90-91
    structure, 83
  evolution of, 9
  fungi, 34
    environments, 36-38
    recycling capabilities, 35-36
  origins of, 25-26
  prokaryotes, compared, 70-71
  protozoa, 26-29
    diseases, 29-30
    environments, 27
    sizes, 26
  role in, 33
    photosynthesis, 30-31
    types, 31-32
  viral infections, 45-46
Euprymna scolopes, 21
exogenous retroviruses, 191-193

Exponential phase (bacterial growth curve), 97
exponents, 98
expression (gene), 67
expulsion, pathogens, 121
extremophiles, 18, 23

F

facultative anaerobes, 15
fastidious microbes, 94
fats, 56
Fausey, Norm, 258
femtoplankton, 255
fermentation, biotechnology, 266-268
Filariasis, 114
filoviruses, 245, 276
filters, sterilization, 137
fission, 97-99
flagellates, 29
flatulence, 163
flaviviruses, 180-184
Fleming, Alexander, 146
flora, normal human flora
  gut microbes, 158-163
  mouth microbes, 166-169
  respiratory tract microbes, 169
  skin microbes, 163-165
  urogenital tract microbes, 170
Florey, Howard, 146
fluconazole, 150
food poisoning, 234-235
food preservation, 240-241
foodborne diseases, 234-235
  Campylobacter jejuni, 235-236
  Clostridium botulinum, 238
  Escherichia coli, 236-237

food preservation, 240-241
Salmonella typhi, 236
Shigella, 237-238
spoilage, 238-240
foot-and-mouth disease, 243
foraminifera, 28
forensics, microbial forensics, 283
forest biotech, 269
fossil fuels, 11
fragmentation, algae, 32
Francisella tularensis, 276
free-living bacteria, 18
fresh water microbes, 256
fungi, 4, 8
  allergens, 37
  antibiotics, 36
  antifungal drugs, 150
  diseases, 35, 37, 226-227
    allergies, 229
    aspergillosis, 230
    asthma, 229
    Blastomyces dermatitidis, 230
    Candida, 227-229
    candidiasis, 227-229
    Coccidioidomycosis, 231
    dermafflictions, 229-230
    Histoplasma capsulatum, 231
    mycotoxins, 229
    Pneumcystis jiroveci, 227
  environments, 36-38
  mildews, 34
  molds, 34
  mushrooms, 34
  mycorrhizae, 35
  penicillin, 36
  recycling capabilities, 35-36
  reproduction, 36

rusts, 35
true fungi, 35
yeast, 34

G

gallstones, 177
gametes, 28
gastritis, 162
gastroenteritis, 219-221, 234-235
  Campylobacter jejuni, 235-236
  Escherichia coli, 236-237
  Salmonella typhi, 236
  Shigella, 237-238
gene expression, 67
gene therapy, 76-78
  viruses, 41
genes, 63
  designer genes, 78-79
  DNA, 64-73
  functions, 65-66
  gene expression, 67
  gene therapy, 76-78
  genetic engineering, 76-79
  genetic sequencing, 73-75
  latent genes, 164
  lateral transfers, 75
  lytic genes, 165
  microbes, swapping without sex, 99-102
  RNA, 67-73
  structural genes, 65
  toxin genes, 75
  transcriptions, 68-71
  translations, 70-71
genetic engineering
  controversies, 79
  with microbes, 76-79
genetics, 64
genetic sequencing, 73-75

genomes, 71-72
  genetic sequencing, 73-75
  Human Genome Project, 73
  metagenome projects, 74
genomics, 64
German measles, 143
germicidal, 133
Giardia, 112
Giardia lamblia, 29, 221
gingivitis, 168
Glanders, 276
glycolysis, 116
gonorrhea, 20, 141, 203
Gram, Hans Christian, 15
gram-negative bacteria, 15-16
gram-positive bacteria, 15-16
green algae, 8, 31
green chlorophylls, 31
green sulfur bacteria, 20
griseofulvin, 150
group A streptococci, 141
group B strep, 141
groupings, bacteria, 19-20
growing microbes, 95-96
guanine, 59
gum disease, caring for, 168
gut microbes, 158-163
gymnosporangium, 113

H

H. pylori, 141, 161-162
Haemophilus, 209
halitosis, 168
halophiles, 23
Hankin, Ernest, 44
Hansen's disease, 211
hantavirus, 183, 244
hantavirus pulmonary syndrome, 183
hantaviruses, 141

heat, sterilization, 134-135
heavy metals, as disinfectants, 138
Helicobacter pylori, 141, 161-162
hemoglobin C, 218
hemoglobin S, 218
hemorrhagic colitis, 244
HEPA (high-efficiency particulate air) filters, 137
hepatitis, 42
hepatitis A, 132, 177-178
hepatitis B, 132, 174, 178-179
hepatitis C, 132, 141, 179
hepatitis D, 179
hepatitis E, 141, 179
hepatovirus, 177-179
herd immunity, 142-143
herpesviruses, 40-42, 141, 149, 164-165, 193-195
heterotrophs, 10, 15-17, 27, 31
high-efficiency particulate air (HEPA) filters, 137
Histoplasma, 248
Histoplasma capsulatum, 231
HIV (human immunodeficiency virus), 67, 123, 130, 141, 175, 190-193
   avoiding, 132
HLA (human leukocyte antigens), 120
homeostasis, 166
hoof-and-mouth disease, 243
hookworm, 112
hopanoids, 11
HTLV (human T cell leukemia viruses), 192
human flora (normal)
   gut microbes, 158-163
   mouth microbes, 166-169
   respiratory tract microbes, 169
   skin microbes, 163-165
   urogenital tract microbes, 170
Human Genome Project, 73
human leukocyte antigens (HLA), 120
human T cell leukemia viruses (HTLV), 192
humoral immunity, 126-128
hydrogen, 52
hydrogen bonds, atoms, 53
hydrogen peroxide, 138
hydroxyl groups, 57
hyphae, 36

I

icosahedrons, 41
ICSP (International Committee on the Systematics of Prokaryotes), 17
IgA constant region (Y-shaped antibodies), 127
IgD constant region (Y-shaped antibodies), 127
IgE constant region (Y-shaped antibodies), 127
IgG constant region (Y-shaped antibodies), 127
IgM constant region (Y-shaped antibodies), 127
immune systems, 119
   active immunity, 285-286
   adaptive immune systems, 120, 124
      cell-mediated immunity, 126
      humoral immunity, 126-128
      lymphocytes, 125-126
      memory, 125, 128
   antigens, 120
   development of, 129
   diseases, struggle against, 120-121
   diseases of, 129-130
   immune-system enhancers, 144
   innate immune systems, 120-121
      barrier methods, 121
      inflammation, 124
      leukocytes, 122-123
      native microbes, 122-124
   MHC (major histocompatibility complex), 120
   passive immunity, 284-285
   responses, bolstering, 283-284
immune-system enhancers, 144
immunoglobulins, humoral immunity, 126-128
immunological memory, 128
immunology, 120
incineration as sterilization, 134
industrial biotechnology, 270-272
infantile paralysis, 176
inflammation, 124
influenza epidemic of 1918, 140
influenza viruses, 41, 184-185, 273
innate immune systems, 120-121
   barrier methods, 121
   inflammation, 124
   leukocytes, 122-123
   microbes, marshalling, 122-124
insect-borne viruses, 180
   bunyaviruses, 183
   dengue fever, 181-182
   rubella, 183
   West Nile virus, 182-183
   yellow fever, 182
interferons, 144

International Committee on the Systematics of Prokaryotes (ICSP), 17
intrinsic resistance, 151
ionic bonds, atoms, 53
ions, 53

J–K

jaundice, 177
Jenner, Edward, 43
Jirovec, Otto, 227
jock itch, 37, 113
junk DNA, 71-72

Kaposi's sarcoma, 191, 194
Karenia brevis, 34
ketoconazole, 150
Korean hemorrhagic fever, 183
kuru, 46

L

La Crosse encephalitis, 183
lactic acid fermentation, 267
lactobacilli, 163
Lactobacillus, 21
Lactobacillus acidophilus, 267
Lactobacillus casei GG, 282
Lactobacillus rhamnosus, 282
Lag phases (bacterial growth curve), 97
lakes, microbes, 256
latent genes, 164
lateral transfer of genes, 75
Leeuwenhoek, Anton von, 20
Legionella, 213
Legionnaires' disease, 30, 213
Leishmania, 224
Leishmania major, 74
leishmaniasis, 74, 141
lentiviruses, 78

leprosy, 211
Leptospira interrogans, 213
leukemias, 191
leukocytes, 122-123
lichens, 31, 34
limestone structures, 28
linezolid, 146
lipids, 56
lipopolysaccharide layer (LPS), 87
livestock biotech, 269
LPS (lipopolysaccharide layer), 87
luciferase, 21
lupus, 129
Lyme disease, 141, 244
lymph, 126
lymph nodes, 126
lymphatic systems, 125-126
lymphocytes, 125-126
    cell-mediated immunity, 126
    humoral immunity, 126-128
lysis, 44
lytic genes, 165

M

macrophages, 123-124
mad cow disease, 10, 47
magnetosomes, 22
Magnetospirillum magnetotacticum, 22
major histocompatibility complex (MHC), 120
malaise, 176
malaria, 29, 113, 141, 216-219
malnutrition, immunological defenses, effects on, 130
Marburg fever, 244-246
marine microbes, 253-255

Marshall, Barry, 162
MDR (multidrug resistance), 151
medicinal biotechnology, 270
medicines
    antibiotics, 145
        combinations, 148
        molds, 146-147
        mycins, 147
        new discoveries, 155-156
        resistance, 150-155
        side effects, 148
        sulfa drugs, 147
        tetracyclines, 147
    antifungal drugs, 150
    antiviral drugs, 149-150
membranes, microbial cells, 84
    archaeal cells, 89-90
    bacterial cells, 85-89
memory, adaptive immune systems, 128
mental retardation, 247
messenger RNA (mRNA), 71
metabolic syndrome, 177
metagenome projects, 74
methanogens, 23
metronidazole, 154
MHC (major histocompatibility complex), 120
microaerophiles, 15
microbes, 3-4, 11
    active transport, 94-95
    adaptation, 7
    aerobes, 10
    algae, 8
    anaerobes, 10
    archaea, 9, 23
    autotrophs, 10
    bacteria, 9
    custom-made microbes, 286-287
    defining, 4

development, 95-96
diffusion, 94-95
diseases, 7
diversity of, 75
elements, 51-54
endosymbiosis, 106
eukaryotes, 8-9
fastidious microbes, 94
fresh water microbes, 256
fungi, 8
genes, swapping without
    sex, 99-102
genetic engineering with,
    76-79
genetically engineering, 76
heterotrophs, 10
human life, 10-12
importance of, 5
lives of, 10
marine microbes, 253-255
mutualism
    biofilms, 106-108
    commensalism, 111-112
    mycorrhizae, 109
    parasites, 112-114
    quorum sensing, 108
    rhizobia, 109-110
normal human flora
    gut microbes, 158-163
    mouth microbes,
        166-169
    respiratory tract
        microbes, 169
    skin microbes, 163-165
    urogenital tract
        microbes, 170
prevalence of, 3, 5-6
prions, 10
prokaryotes, 6
    archaea, 13
    bacteria, 13-22
    evolution of, 9
protozoa, 8
reproduction, 97-102

required nutrients, 93-95
scientific research, 7-8
sewage, 256-259
soil, 259-264
    biological soil crusts,
        262
    compost, 263-264
symbiosis, 106
types, 4, 8
viruses, 10
microbial cells
    archaeal cells, 89-90
    bacterial cells, 85-89
    commonalities, 82-84
    eukaryote cells, 90-91
    membranes, 84
    structure, 82-85
    walls, 84
microbial forensics, 283
microbiologists, 5
microbiology, 4
    careers, 287-288
    future of, 280
Micrococcus, 16
microorganisms, 4
mildews, 34
mitochondria, 27, 115-117
MMR vaccine, 143
molds, 34
    antibiotics, 146-147
    water molds, 38
molecules, 52-53
    amino acids, 56
    carboxylic acid, 57
    organic molecules, 54
        carbohydrates, 55
        lipids, 56
        nucleic acids, 58-61
        proteins, 56-60
    water, 53-54
mononucleosis, 194
monosaccharides, 55
mouth microbes, 166-169

mRNA (messenger RNA), 71
MRSA (methicillin-resistant
    Staphylococcus aureus), 153
mucous, 121
multidrug resistance (MDR),
    151
multiple sclerosis, 129, 247
mushrooms, 34
mutations, DNA, 66
mutualism
    biofilms, 106-108
    commensalism, 111-112
    mycorrhizae, 109
    parasites, 112-114
    quorum sensing, 108
    rhizobia, 109-110
mycelium, 36
mycins, 147
Mycobacterium, 210-211
Mycobacterium tuberculosis, 123
mycology, 34
Mycoplasmas, 17, 207
mycorrhizae, 35, 109
mycotoxins, 229
myxobacteria, 19
myxoma virus, 175

N

naftifine, 150
nanobacteria, 17
nanoplankton, 255
nasopharyngeal carcinoma,
    195
native microbes, innate
    immune systems, 122-124
Neisseria, 16, 166
Neisseria gonorrhoeae, 203
neutrons, 52
neutrophils, 122-124
nitrogen, 52
non-A, non-B hepatitis, 179
nori, 31

normal anaerobic flora, diseases from, 205-206
normal human flora
  diseases from, 205-206
  gut microbes, 158-163
  mouth microbes, 166-169
  respiratory tract microbes, 169
  skin microbes, 163-165
  urogenital tract microbes, 170
noroviruses, 190, 245
novelty-seeking, 226
nuclei, protozoa, 27
nucleic acids, 58-61
nucleoids, 18
nucleotides, 59
nutrients, microbes, 93-95
nystatin, 150

O

obligate intracellular parasites, 39
obsessive-compulsive disorder, 247
oceans, marine microbes, 253-255
oligosaccharides, 55
oomycetes, 38
organelles, 114-118
organic molecules, 54
  carbohydrates, 55
  lipids, 56
  nucleic acids, 58-61
  proteins, 56-60
Orthomyxovirus, 184
osmosis, bacterial cell walls, 85
oxygen, 52
  bacteria, relationship with, 15

P

P. carinii, 227
P. meibomiae, 242
papillomaviruses, 174, 193
parainfluenza, 141, 175
Paramecium, 29
paramyxoviruses, 186
parasites, 112-114
  obligate intracellular parasites, 39
  protozoa, 29
particles, 52
parvovirus, 187-188
passive immunity, 142, 284-285
Pasteur, Louis, 268
pathogens, 7
  animals, 242-243
  avoiding, 131-132
  destroying, 132-133
    antiseptics, 139-140
    disinfectants, 137-139
    sterilization, 133-137
  plants, 241-242
PCR (polymerase chain reaction), 21
Pelagibacter ubique, 254
penicillin, 36, 133, 146
Penicillium, 268
peptide bonds, 57
peptidoglycan (PGN), 85-87
periodontitis, 168
peritonitis, 158
pertussis, 210
petroleum, 11
Pfiesteria piscicida, 34
PGN (peptidoglycan), 85-87
pH (potential of hydrogen), 96
phages, 44-45
phagocytes, 122-123

phagocytosis, 123
Phakopsora pachyrhizi, 242
pharmaceuticals
  antibiotics, 145
    combinations, 148
    molds, 146-147
    mycins, 147
    new discoveries, 155-156
    resistance, 150-155
    side effects, 148
    sulfa drugs, 147
    tetracyclines, 147
  antifungal drugs, 150
  antiviral drugs, 149-150
  medicinal biotechnology, 270
phenol, 139
phenolics, 139
pheromones, 19
phosphatidyl ethanolamine, 56
photosynthesis, 4, 8-9, 15
  algae, 30-31
phycoerythrin, 31
Phytophthora infestans, 38
phytoplankton, 31
picoplankton, 255
picornaviruses, 175
  enteroviruses, 176
  hepatovirus, 177-179
  rhinoviruses, 175-176
pili, 18
plague, 20, 208-209, 244, 276
Plague of Justinian, 209
plankton, 255
plants, pathogens, 241-242
plaques, 124, 166
plasma, 126
plasmids, bacteria, 18
Plasmodium, 217
plastids, 31
Pneumocystis jiroveci, 227

pneumocystis pneumonia, 35, 191
pneumonia, 20, 42
pneumonic plague, 209
  biological warfare, 274
polio, 43, 67
poliovirus, 176
polymerase chain reaction (PCR), 21
polymorphonuclear leuko-cyte, 122
polymyxin, 147
polypeptides, 58
polysaccharides, 55, 85-86
ponds, microbes, 256
Porphyromonas gingivalis, 168
potato blight, 38
poxviruses, 40, 195
prebiotics, 281
preservation, food, 240-241
prevention, diseases, 131-133, 143-144
  antiseptics, 139-140
  disinfectants, 137-139
  sterilization, 133-137
  vaccines, 140-143
Priest Pot, 256
primary structures, proteins, 57
prions, 10
  diseases, 46-47
  mad cow disease, 47
  transmissible spongiform encephalopathies (TSE), 48
probiotics, 281-283
Prochlorococcus, 254
prokaryotes, 6, 25
  archaea, 13
  bacteria, 13-22
  cells, structure, 82
  eukaryotes, compared, 70-71
  evolution of, 9

Propionibacterium, 21
Propionibacterium acnes, 164
protein folding, 58
protein subunits, 57
proteins, 56-60
  abnormal protein, 66
  amino acids, 57
  cells, 65
  DNA, 67-73
  protein folding, 58
  protein subunits, 57
  regulatory proteins, 66
  RNA, 67-73
  structures, 57
protons, 52
protozoa, 4, 8
  African sleeping sickness, 29
  amoebas, 29
  ciliates, 29
  classifications, 29
  cysts, 28
  diseases, 28-30
  environments, 27
  flagellates, 29
  nuclei, 27
  parasites, 29
  reproduction, 28
  sizes, 26
  specializations, 27
  sporozoans, 29
protozoan diseases, 215, 225-226
  gastroenteritis, 219-221
  Leishmania, 224
  malaria, 216-219
  Toxoplasma gondii, 225
  Trichomonas vaginalis, 221-222
  Trypanosoma brucei, 223-224
  Trypanosoma cruzi, 222-223
PrPCs, 47
Prusiner, Stanley, 47

Pseudomonas, 141, 207-208
Pseudomonas carboxydovorans, 21
psittacosis, 276
psoriasis, 129
pyramids of Egypt, 28

Q–R

Q fever, 276
quarks, 52
quaternary structures, proteins, 57
quorum sensing, 19-20, 108

rabies, 132, 141, 184
radiation, sterilization, 135-136
recycling
  fungi, 35-36
  microbial role in, 6
red algae, 8, 31
Red tides, 33
regulatory proteins, 66
reproduction
  bacteria, 18
  fungi, 36
  microbes, 97-102
  protozoa, 28
resident microbes
  gut microbes, 158-163
  mouth microbes, 166-169
  respiratory tract microbes, 169
  skin microbes, 163-165
  urogenital tract microbes, 170
resistance, biotics, 150-155
respiratory syncytial virus (RSV), 141, 175
respiratory tract microbes, 169
Reston strain (Ebola), 246

retroviruses, 78
  endogenous retroviruses, 190-191
  exogenous retroviruses, 191-193
reverse transcription, 190
rhinoviruses, 175-176
rhizobia, 109-110
Rhodospirillum rubrum, 256
ribosomal RNA (rRNA), 71
ribosomes, 67
  translations, 69-70
Rickettsia, 18, 115
Rickettsia prowazekii, 276
Rickettsiae, 213
Rift Valley fever, 183
rimantadine, 149
ringworm, 113, 229-230
rivers, microbes, 256
RNA (ribonucleic acid), 59-61
  bases, 67
  DNA, compared, 67-73
  proteins, 67-73
  translations, 69-70
  viruses, 64, 67
RNA polymerase (RNAP), 68
Rocky Mountain spotted fever, 115, 213
rotavirus, 188-189
roundworm, 112
rRNA (ribosomal RNA), 71
RSV (respiratory syncytial virus), 175
rubella, 142, 183
rusts, 35

S

S. aureus, 141, 152
S. mitis, 21
Saccharomyces cerevisiae, 36, 101-102
Salmonella, 124, 141, 220, 276
Salmonella typhi, 214, 236

San Joaquin fever, 231
SAR11, 254-255
Sarcina, 16
sarcomas, 191
SARS (severe acute respiratory syndrome), 67, 186-187
schistosomiasis, 141
schizophrenia, 247
SCID (severe combined immunodeficiency), 78, 129
scientific research
  archaea, 23
  microbes, 7-8
scrapie, 47
secondary endosymbiosis, 117
secondary structures, proteins, 57
septic shock, 124
serum hepatitis, 178
severe immune deficiency (SCID), 78
sewage, microbes, 256-259
shapes
  archaea, 23
  bacteria, 16
Shigella, 141, 220, 237-238, 276
Shigella dysenteriae, 75
shingles, 175
side effects, antibiotics, 148
Sin Nombre virus, 183
skin, fungal diseases, 229-230
skin microbes, 163-165
skin, as immunity barrier, 121
slow viruses, 47
smallpox, 10, 40, 276
  bioterrorism, 274-275
  eradication of, 43
soil, microbes, 259-264
  biological soil crusts, 262
  compost, 263-264
soybean rust, 242
specializations, protozoa, 27
spirilla, 16

spoilage, food, 238-240
spores, algae, 32
sporozoans, 29, 217
St. Louis encephalitis virus, 182
Stanford University project, endogenous intestinal microflora, study of, 160-161
staph, 153
Staphylococcal enterotoxin B, 276
Staphylococcus, 16, 146, 199-200, 273
Staphylococcus aureus, 164, 244
Staphylococcus epidermis, 164
Stationary phase (bacterial growth curve), 98
sterilization, 132-133
  chemicals, 136
  filters, 137
  heat, 134-135
  radiation, 135-136
stomach flu
  norovirus, 190
  rotavirus, 188-189
stomach microbes, 158-163
streams, microbes, 256
streptococci, 166
Streptococcus, 16, 45, 146, 201-202, 273
Streptococcus agalactiae, 247
Streptococcus mitis, 18
Streptococcus mutans, 167
Streptococcus pneumoniae, 147, 151
Streptococcus thermophilus, 21
Streptomyces, 150, 268
streptomycin, 147
structural genes, 65
structures
  microbial cells, 82-85
  proteins, 57
subatomic particles, 52

subclinical viruses, 174
sugar glucose, 55
sulfa drugs, 133, 146-147
surfactants, 138
Surveyor 3, 18
symbiosis, 9, 106
Synechococcus, 254-255
synthetic biology, 286-287
syphilis, 141

T

T cells, 125-126
Tamiflu, 149
temperate phages, 44
temperature, microbe development, 95
terrorism, bioterrorism, 273-277
tertiary structures, proteins, 57
tetracyclines, 147
thermoacidophiles, 23
Thermus aquaticus, 21
Thiomargarita namibiensis, 16
thymine, 59
tobacco mosaic virus (TMV), 40
toenail fungus, 35
togaviruses, 180-184
tooth decay, 20, 166
    caring for, 167-168
toxin genes, 75
Toxoplasma gondii, 225
transcriptions, 68-71
transduction (DNA acquisitions), 100
transfer RNA (tRNA), 71
transformations (DNA acquisitions), 99
translations, 70-71
    ribosomes, 69-70

transmissible spongiform encephalopathies (TSE), 48
transposons, 72-73, 151
Treponema pallidum, 212-213
Trichomonas vaginalis, 29, 221-222
tRNA (transfer RNA), 71
true fungi, 35
Trypanosoma brucei, 74, 223-224
Trypanosoma cruzi, 74, 222-223
trypanosomes, 29
TSE (transmissible spongiform encephalopathies), 48
tuberculosis (TB), 20, 141, 211
Tularemia, 274, 276
Tumpey, Terrence, 5
type I diabetes, 129
typhoid fever, 214
typhus, 214, 276

U-V

ulcers, 20
uracil, 59
urogenital tract microbes, 170

vaccinations, 132
    edible vaccinations, 285-286
    herd immunity, 142-143
    MMR vaccine, 143
    viruses, 42
vaccinia, 40
vancomycin, 152-154
vancomycin-resistant enterococci (VRE), 152
variant CJD (vCJD), 48
Varicella-Zoster virus, 194, 247
variola virus, 43
vectors, 18

vegetation, diseases, 241-242
venereal warts, 247
Verticillium, 261
Vibrio cholerae, 44, 244, 277
Vibrio fischeri, 21
Vibrio harveyi, 21
vibrios, 16
vinegar, 267
virions (viruses), 40
viroids, 46
virology, 40
virtual microbial cells, 286
virulent phages, 44
viruses, 4, 10, 25, 39-40, 173-175
    AAV (adeno-associated virus), 78
    adenoviruses, 41, 78, 187
    antiviral drugs, 149-150
    arenaviruses, 183
    avoiding, 131-132
    bacteria, infection of, 43-44
    capsids, 40
    carcinogenic viruses
        herpesviruses, 193-195
        papillomaviruses, 193
    classifications, 40
    coronaviruses, 186-187
    cowpox virus, 40
    discovery of, 40-42
    diseases, 40-46
    E. coli, 41
    Ebola, 245-246
    envelopes, 40
    eukaryotes, infections of, 45-46
    gene therapy, 41
    hepatovirus, 177-179
    herpesvirus, 40
    icosahedrons, 41
    influenza viruses, 41, 184-185

insect-borne viruses, 180
  bunyaviruses, 183
  dengue fever, 181-182
  rubella, 183
  West Nile virus,
    182-183
  yellow fever, 182
Marburg fever, 245-246
norovirus, 190
origins of, 42
paramyxoviruses, 186
parvovirus, 187-188
phages, 44-45
picornaviruses
  enteroviruses, 176
  rhinoviruses, 175-176
poxviruses, 40, 195
rabies, 184
retroviruses, 78
  endogenous retro-
    viruses, 190-191
  exogenous retroviruses,
    191-193
RNA, 64, 67
rotavirus, 188-189
shapes, 41
slow viruses, 47
smallpox, eradication of, 43
subclinical viruses, 174
vaccinations, 42
vaccinia, 40
virions, 40
viroids, 46
VRE (vancomycin-resistant
  enterococci), 152

W–X

Waksman, Selman, 147
walls, microbial cells, 84
  archaeal cells, 89-90
  bacterial cells, 85-89
  eukaryote cells, 90-91
warts, 42
water, 53-54
water molds, 38
water treatments, 257-259
weapons, bioweapons,
  274-277
West Nile virus, 67, 182-183,
  245, 273
white blood cells
  leukocytes, 122-123
  lymphocytes, 125-126
white cliffs of Dover, 28
whooping cough, 210
Wigglesworthia, 110
Wilson's disease, 177
wine, production of, 21
Wolbachia, 113, 225

Y–Z

Y-shaped antibodies, 126-127
yeast, 34-36
  infections, 37
  reproduction, 36
yellow fever, 42, 132, 182
Yersinia pestis, 208-209, 276
yogurt, production of, 21

Zymonas, 22